GENERAL SCIENCE FOR THE BAHAMAS
GRADE 8

For Students Studying for

The Bahamas Junior Certificate (BJC) Examination

Jacquelyn K. Argyle

B.Sc. (Biology), MA (Curriculum and Instruction)

This book is dedicated to my very loving husband Sean Anthony Argyle.

Acknowledgements

Teaching Science specifically for the Bahamas Junior Certificate (BJC) Examination involves much research. One of my goals for writing this book is to provide accurate and relevant content information for both teachers and students. For this book, I thank all persons who impacted my life in any way, especially those who motivated me to begin this writing venture. With great appreciation, I acknowledge my husband Sean Argyle, for the cover design and his help, love and motivation during the writing process.

I thank my siblings who encouraged me to publish this book and for their constant love and support during the many hours of writing. I owe a hearty thank you to Mr. Lee M. Van Lew Sr. for the many long hours he spent proofreading; Mrs. Evelyn Lopez for her editing expertise; my niece Jaide Williams, for her skillful assistance with typing, my nephew Sekeron Dorsette, for his striking graphic designs. Last, but certainly not least, Almighty God, for with Him, I can do all things.

Table of Contents

Introduction

This book is intended for teachers and students of General Science Grade 8 studying for the Bahamas Junior Certificate Examination (BJC). The goal of this book is to cover the required topics in a comprehensive manner. The content was compiled from Ministry of Education Curriculum Guidelines, Grades – 7, 8 and 9- Department of Education 2010.

Key words are shown in bold type and diagrams have been used whenever possible. Comprehension questions have been included at the end of each topic to enable students and teachers to assess knowledge and understanding of the topic. Some field activities have been included as well.

Science is interesting and exciting. It will give students knowledge and skills they will use throughout life. It will provide opportunities to engage and expose them in acquiring scientific knowledge to enhance their critical thinking skills. Science will help students to understand the world in which they live. For example, science explains how airplanes fly and how birds find their way when they travel long distances. In addition, people use science to make discoveries that have practical value. One of these discoveries is the use of light to carry telephone messages through a glass wire. Another is the lengthening of human life through the use of heart pacemakers and other mechanical devices.

A science class provides great opportunities to increase a student's understanding of the world. It allows them to study the environment and learn about ways in which they can play an active role in conservation.

In the future, there will be an endless number of new scientific discoveries. The discoveries will affect the career and daily life of our students. It is my hope that students would be interested in science to the extent that they may pursue a career in science and make contributions to more discoveries.

Taxonomy - Classification of Living Things

Taxonomy is a system for classifying and identifying organisms according to their similarities and differences. This system was developed by Swedish scientist Carl Linnaeus in the 18th Century.

Binomial Nomenclature

Linnaeus's taxonomy system has two main features that contribute to its ease of use in naming and grouping organisms. The first is the use of binomial nomenclature.

The system of giving a scientific name to each properly identified plant or animal is called **nomenclature.** A system of nomenclature of plants and animals in which each scientific name consists of two parts or sub-names is called the system of **binomial nomenclature.** This means that an organism's scientific name is made up of a combination of two terms. These terms are the **genus** name and the **species**. Both of these terms are italicized and the first letter of the genus name is also capitalized. For example, the scientific name for humans is *Homo sapiens*. The genus name is *Homo* and the species is *sapiens*. These terms are unique and no other species can have this same name.

Classification Divisions (Categories)

The second feature of Linnaeus' taxonomy system that simplifies organism classification is the ordering of species into seven broad divisions. The divisions are as follows:

- Kingdom
  - Phylum
    - Class
      - Order
        - Family
          - Genus
            - Species

A good aid for remembering these categories is the mnemonic device:

Keep Plates Clean Or Family Gets Sick or King Philip Came Over For Good Soup

Classification levels are more specific at the bottom. For example, all animals belong to the animal kingdom; fewer belong to the same phylum, and even fewer belong to the same family. Species however, is the most specific classification. Species may be defined as a group of organisms that closely resemble each other physically, behaviorally and internally. They usually mate among themselves and produce fertile offspring. Features of some species in a genus are very similar. The members of each group have certain features in common which distinguish them from other groups. However breeding will not produce fertile offspring.

This system follows certain rules, such as:

1. The scientific name must be in Latin.
2. Genetic name should come first and must begin with a capital letter.
3. The same name should not be used for two or more species under the same genus.
4. The scientific name must be written in italics.

Below is a list of organisms and their classification in this taxonomy system using the major categories.

Taxonomy of Organisms

	Brown Bear (A)	House Cat (B)	Dog (C)	Killer Whale (D)	Wolf (E)
Kingdom	Animalia	Animalia	Animalia	Animalia	Animalia
Phylum	Chordata	Chordata	Chordata	Chordata	Chordata
Class	Mammalia	Mammalia	Mammalia	Mammalia	Mammalia
Order	Carnivora	Carnivora	Carnivora	Cetacea	Carnivora
Family	Ursidae	Felidae	Canidae	Delphinidae	Canidae
Genus	*Ursus*	*Felis*	*Canis*	*Orcinus*	*Canis*
Species	*Ursus arctos*	*Felis catus*	*Canis familiaris*	*Orcinus orca*	*Canis lupus*

Comprehension Questions

1. Define the term taxonomy
2. Who is responsible for the present day system of classification?
3. List the 7 divisions of classification in order.
4. What is the highest taxonomic rank?
5. What is the most specific taxonomic rank?
6. Define the term species.
7. What do you understand by the term binomial?
8. What TWO parts make up the scientific name?
9. Which part of the scientific name is capitalized?
10. Explain TWO reasons why scientific names are important.
11. Examine the classification of the animals listed in the chart and answer the questions.
a. Compare the Brown Bear and the wolf, how many groups are the same?
b. Compare the house cat and the killer whale, how many groups are the same?
c. Which two animals (A and B, A and C or C and E) are most alike in classification?
d. Which two animals from the chart have most of the same traits?
e. Which two animals are most closely related?
f. Which two animals look more alike?
g. What is the scientific name of brown bear?
h. What does the first word in the scientific name represent?
i. What does the second word in the scientific name represent?
j. A naming system that gives every organism a two-word name is _____.

Dichotomous Key (Tool for Identifying Organisms)

A dichotomous (di- kah- tuh- mus) key is a tool used to identify unknown organisms. Using this tool is one way you and scientists can solve problems scientifically.

A dichotomous key is a device that can be used to easily identify unknown plants and animals. The word dichotomous comes from two Greek words that together mean, "Divided in two parts". A dichotomous key consists of a series of two part statements that describe characteristics of organisms. At each step of a dichotomous key the user is presented with two choices, yes or no. If you answer each question correctly, the key will lead you to the name of the plant or animal. If you answer questions incorrectly, the key will tell you to start over at question 1 so that you can answer the questions more carefully. Eventually the user will be led to the name of the organism that they are trying to identify. Below is a simple dichotomous key. Follow the steps to name the creatures.

If it is true,	do this.
1. The creature has two eyes.	Go to step 2.
2. The creature has one eye.	Go to step 5.
3. The creature has one or more antennae.	Go to step 3.
4. The creature has no antennae.	Its name is "L."
5. The creature has one antennae.	Its name is "I."
6. The creature has more than one antennae.	Go to step 4.
7. The creature has two antennae.	Its name is "S."
8. The creature has three antennae.	Its name is "Y."
9. The creature has one or more antennae.	Go to step 6.
10. The creature has no antennae.	Its name is "A."
11. The creature has one antennae.	Its name is "F."
12. The creature has two antennae.	Its name is "C."

Land Crab - Cardisoma guanhumi

The land crabs, *Cardisoma guanhumi* are terrestrial. They live on land but they return to the sea only to drink and breed. They are also known as Blue Land Crab, White Land Crab or Giant Land Crab.

The crab's body is protected by an exoskeleton consisting of a smooth carapace made up of a cephalothorax and abdomen. The exoskeleton both protects them from predators and provides support. Connected to its cephalothorax are five pairs of appendages (legs) covered with tactile setae (hair). The front two legs of the crab are the closest to its head. They are sometimes used for walking but mostly used for eating. Those two legs are called chelipeds. At its front are two pairs of antennae and two pincers, one larger than the other. They have a flattened body and two eyes located at the ends of stalks.

The males are generally larger than females. Juveniles are usually tan or brown in color with orange legs. The adult colours range from blue to violet and some females are white or ashy gray.

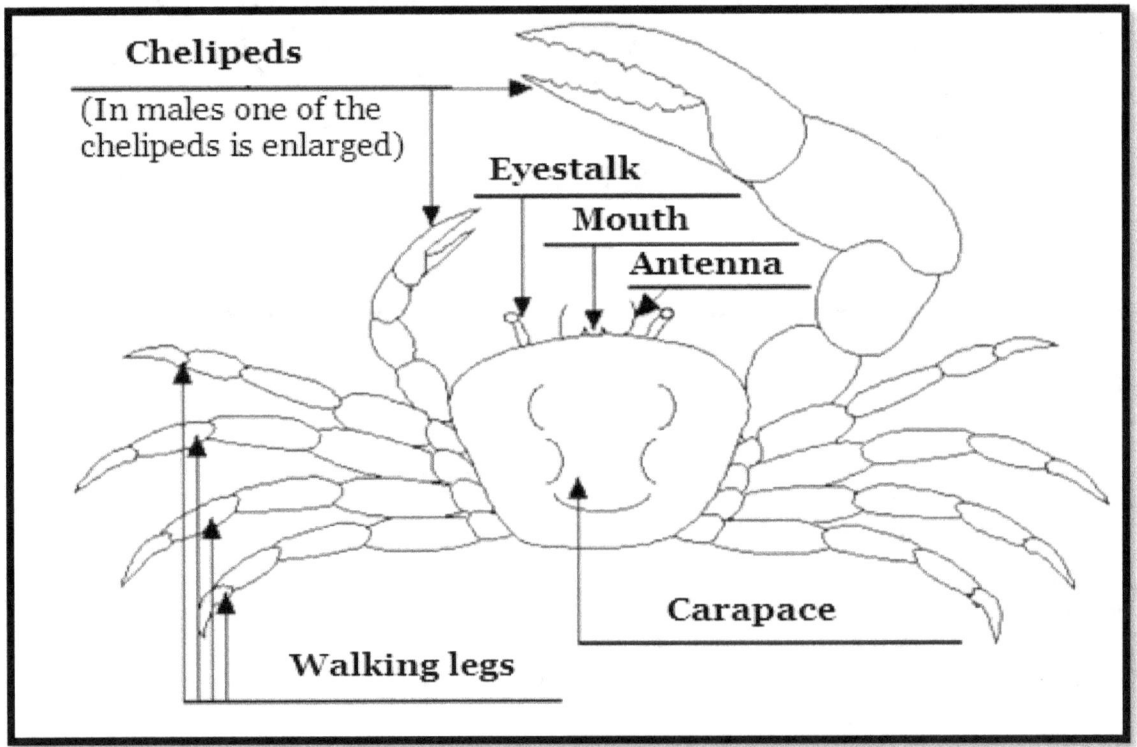

The diagram shows the external features of the crab

Parts of the Crab

Carapace - The hard, protective outer shell of the crab. The carapace is made of chitin.
Eyestalk - The support for the eye.
Mouth - The mouth is located at the front of the crab, near the base of the eyestalks and the antennae.

Walking legs - Four pairs of long, jointed legs used for locomotion (walking). Crabs walk sideways.

Cheliped - Much bigger than walking legs. Chelipeds can also be called claws or pinchers. They are used for defense and food handling. In male crabs, one cheliped is much bigger than the other; in females, the two chelipeds are about the same size.

Tactile Setae- Hairs on the body that are used as touch receptors in crabs.

Habitat

Cardisoma guanhumi lives within several hundred meters of the shore. The crab spends most of its time in its burrow in mud, or coastal sand above the tide line, when it is not foraging or migrating to mate and spawn. Their burrows extend downward until they reach the water table. They can tolerate both freshwater and saltwater conditions, which is why they are often found inland. The crabs crawl down their burrow and periodically submerge themselves in the water to help them maintain gill function and body moisture. Some insects and other small arthropods may be found in these burrows.

Gender of the Land Crab

In adults, the sexes can be distinguished by their abdominal shape. The female abdomen widens and becomes triangular in shape and then semi-elliptical. The male's abdomen is narrower. Adult males are generally larger than females and often have a larger major cheliped

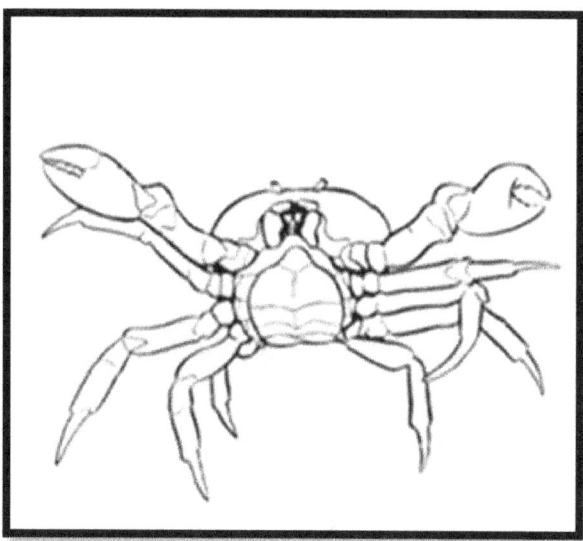

Male – Narrow Abdomen   **Female** – Wide Abdomen

Comprehension Questions

1. What is the scientific name of the land crab?
2. What do you understand by the term terrestrial?
3. State two reasons why crabs return to the sea.
4. Tell if the crab is a vertebrate or invertebrate. Give reasons for your answer.
5. The _____ covers the head and thorax.
6. How many legs does the crab have?
7. What covers the legs of the crab?
8. Explain two functions of the front legs of the crab.
9. Where do crabs spend most of their time?
10. How can you distinguish the gender of the crab?

Reproduction and Development of the Land Crab

The life cycle of the Cardisoma guanhumi (land crab) begins during the rainy season. Their reproductive cycle is heavily dependent on weather. They reach sexual maturity at about four years of age. Spawning, which is a form of reproduction, occurs from June – November or when spring rains begin. Fertilization is internal; once the eggs are fertilized, during a full moon, crabs return to the ocean to deposit their eggs. Gravid (egg-bearing) females migrate to the water's edge and release over 250,000 eggs.

In preparation for their migrations, females gain substantial weight. Following copulation, females carry their eggs for approximately two weeks. Between 20,000 and 1,200,000 eggs are fertilized. At this point, the eggs begin to hatch, and she shakes them off into the ocean. Migrating crabs must avoid natural predators and automobiles as they cross the roadways.

Life Cycle of the Crab

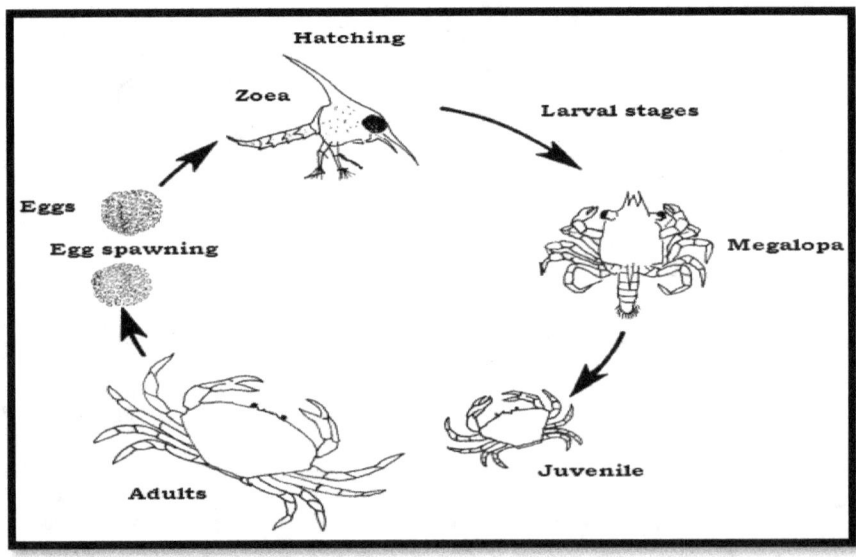

The diagram shows the life cycle of the crab

Life Cycle of a Growing Crab

When the baby crabs hatch, they are called larvae. They do not look like the adult crabs. They have to go through stages of their life cycle before they look like crabs.

The first stage is known as a zoea. When the crab is a zoea, it is usually about 1 millimeter (1/ 25 inches) long. The zoea usually feeds only on the larvae of oysters and starfish. The zoea starts to shed the outer layer of its shell because it is getting big. This is called molting.

The second stage of growth is called megalops. At this stage it has large eyes and gets its claws and the appendages of an adult crab.

After the megalops, the crab enters the juvenile stage of growth. The juvenile crab has all of the body parts that an adult crab has. The hard part that is made during this process is called a chitin. A chitin is the hard part of the crab which is also known as a carapace. The puberty molt comes just before full sexual maturity. Molting regulates the life cycle of the crab.

Range

Cardisoma guanhumi may be found in tropical and sub-topical estuaries and other maritime areas of land along the Atlantic coast of the Americas, from Brazil and Colombia, through the Caribbean, to the Bahamas, and as far north as Vero Beach, Florida.

Growth and Colours

As the crab grows, its exoskeleton does not, so it must molt its old exoskeleton in order to house its expanding body. To prepare for molting, the tissue layer under the exoskeleton detaches and secretes a new exoskeleton below the hard outer one. When the new exoskeleton is completely formed, the old exoskeleton splits along weak points and the animal pulls out, leaving its old exoskeleton intact except for the split. During molting, the crab stops eating and seeks shelter in order to avoid predation. During this process the crab is highly vulnerable to predators because their shell is soft and can be easily penetrated by predators.

Juveniles develop one cheliped (claw) larger than the other, but as juveniles, the size and shape does not distinguish males from females. They are usually tan or brown with orange legs and a purple carapace.

Adult colors range from blue to violet but may also be gray or tan. Spawning females tend to be gray or tan and return to blue after the spawning season

Food Habits

Cardisoma guanhumi is an omnivore. It feeds on plant and animal matter. Although it prefers leaves, fruits, flowers, berries and grasses, C. guanhumi also feeds on insects like spiders, carrion (rotting flesh of dead animals), and faeces. To forage, it typically does not stray far from its burrow and uses light and sound to find food. After foraging, it carries its food in its claws back to its burrow, eats, and saves whatever it does not finish for later.

Predation

Crabs are eaten by large birds, mammals and other crabs but humans are the largest predators of the crab. Humans harvest giant land crabs in massive quantities for food. Jellyfish and small fishes feed on larval crabs, while large mammals (wild hogs) and wading birds feed on juveniles and adult crabs.

Importance for Humans

Cardisoma guanhumi is a significant source of food in various parts of the Caribbean, particularly in the Bahamas. Crabs can be found on many of the Bahama Islands, but they are in abundance on the island of Andros. Andros is known for its Annual Crab Fest. There is a day of celebration and crabs are cooked and sold in many different ways. The crabs are caught, caged and shipped to the capital and other islands to be sold. It is done mainly by roadside vendors. An example of a roadside cage is shown below.

Crabs in a pen

Comprehension Questions

1. Draw and label the parts of the crab. Use colored pencils or crayons and colour the crab with the juvenile colours.
2. What is the scientific name of the land crab _____
3. Explain the growth process of the land crab.
4. State why crabs are vulnerable after molting.
5. How can you distinguish between the juvenile and adult crabs?
6. Rewrite the stages of development of the crab in the correct order from the youngest to the oldest:

Zoea   Megalops   Adult   Egg   Juvenile

7. Crabs are described as omnivores. Explain this term and list THREE organisms that crabs feed on. 8. List THREE predators of the crab.
9. Explain TWO reasons why crabs are important to the Bahamas.

Relationship between Mating and the Rainy Season

Mating begins during the rainy season. Reproductive cycle heavily depends on the rainy season; therefore a major decrease in rainfall will result in a decline in the population.

Ecological and Economic significance

While some see land crabs as a nuisance because of their burrows, they are a vital part of many Bahamian and Caribbean ecosystems and diets.

Crabs are an important food source throughout the Bahamas and the Caribbean. Populations may be in decline in these areas due to overharvesting, land-clearing and development.

Reproduction and the predictability of behavior i.e. movement toward the sea to spawn during the rainy season could lead to over-harvesting and a decline in numbers.

Females carry eggs that can number from 20,000-1,200,000. However, many giant land crabs do not survive the larval stage as a result of predators and other factors. Therefore, many off springs are produced in comparison to those that survive to full maturity.

The crab is not endangered, but there are concerns about the harvesting as they have been captured on a large scale for food in the Caribbean.

No restriction on crab catch will result in over-harvesting and a decline in the crab population in the very near future. A decline in the crab population will result in a demand for crab meals which in turn will result in an increase in the rate of catching crabs. The higher the demand for crabs the more expensive the crab becomes.

How you can help

- Protect land crab burrows from erosion and fill-in from development.
- Be careful of migrating crabs in roadway during spawning / migration season (June-November)
- Do not capture egg bearing females
- Be aware of the regulations

Comprehension Questions

Fill in the blanks with the correct terms.

1. Crabs grow through a process called _____
2. This is what forms under the hard shell. _____
3. The crab seeks shelter to avoid _____
4. The new shell is _____ and can easily be penetrated by _____ .
5. Legs of juvenile crabs are _____ .
6. Adult colours may be _____ , _____ , _____ or _____ .
7. Spawning females are normally _____ or _____ .
8. Cardisoma guanhumi feeds on plant and animals. The term used for this type of feeding is

_____
9. The crab carries its food in its _____ .
10. Three predators of the land crab are _____ , _____ and

_____ .
11. Mating in land crabs begins during the _____ .
12. A decrease in rainfall will result in the _____ in the population of crabs.
13. Populations of crabs may decline due to _____ ,

_____ and _____ .
14. Many land crabs do not survive to maturity because of _____ .
15. No restriction on crab catch will result in _____ and a

_____ in the crab population.
16. A decline in the crab population will result in a _____ for crab meals.
17. The higher the demand the more _____ the crab becomes.
18. The crab is not endangered, but you should not capture _____ females.
19. You should protect their _____ from erosion and fill from development.
20. Write a short story to persuade residents of the importance in protecting the habitat of the land crab.

Activities
1. Use a ruler to measure the width of a model crab.
2. Use measurements to draw a crab 50% of its size.
3. Use a balance to measure the weight of a crab.
4. Make a model of a land crab.
5. Design a pen for land crabs.
6. Make an oral presentation (with visual aids) to describe the life history of the crab.
7. Make an oral presentation with visual aids to describe the habitat of the crab.
8. Participate in a debate on the importance of implementing and observing the closed season for catching crab

The Nassau Grouper

Genus: *Epinephelus*
Species: *striatus*

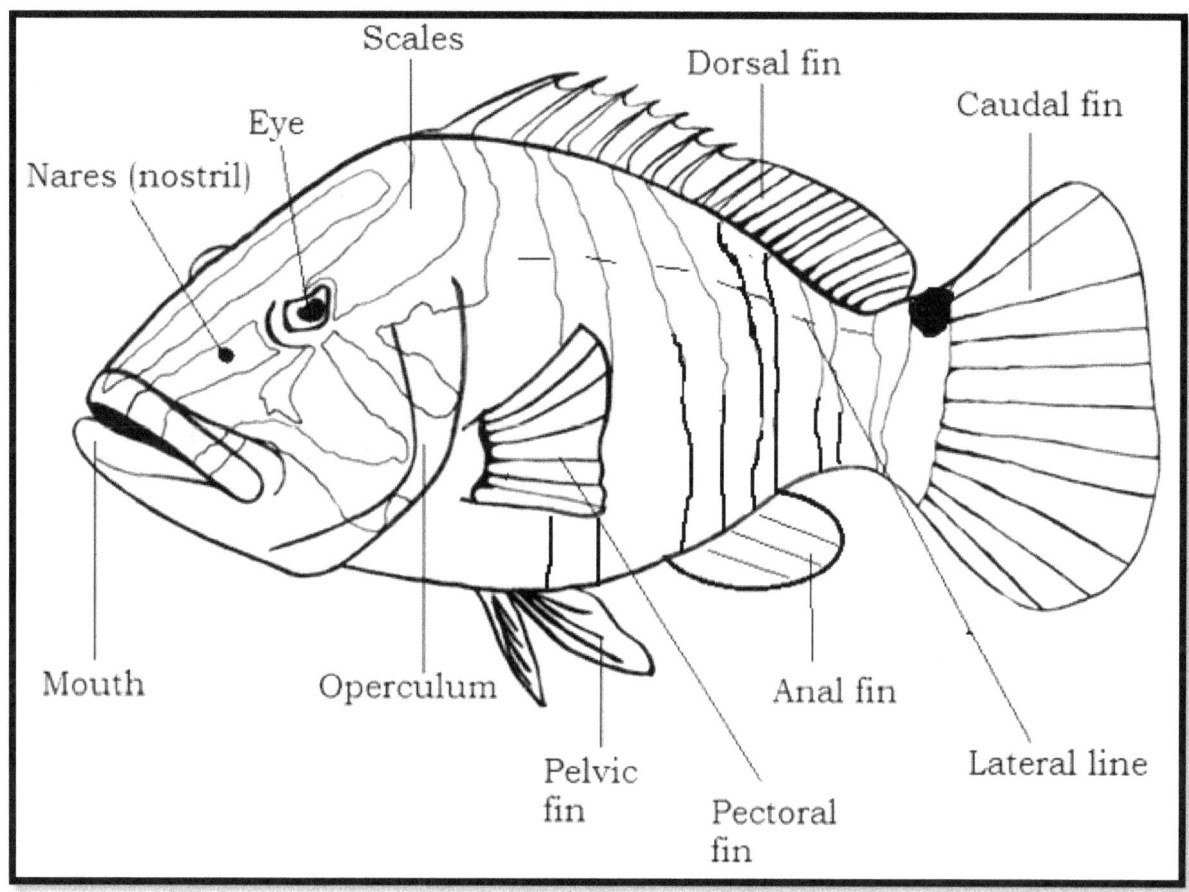

Parts & Function of the Nassau grouper

Part	Description / Function
Eyes	The organs of sight that are found on the head.
Fins	Extend from the body for propelling, steering or balancing in the water, dorsal, caudal, anal, pectoral and pelvic. Pectoral and pelvic and paired fins.
Gills	Below the operculum, used for breathing.
Lateral line	Sensory pores used to detect movement in the water.
Nares (nostril)	Used for smelling.
Operculum	Also called gill cover is a bony flap that protects gills.
Scales	Are small, thin plates that provide protection for the body. They also allow flexibility.

Description of the Nassau Grouper

The Nassau grouper's scientific name is Epinephelus striatus. It is a saltwater fish found in the tropics.

Distinguishing Marks of the Nassau Grouper

The grouper family, serranadae, is very large in size and includes the yellow fin grouper, tiger grouper, black grouper, rock hind, Jew fish gag and red grouper among many others.

The Nassau grouper may be distinguished from other groupers because it has five dark brown vertical bands around its body, a black saddle-like patch near its tail fin, a dark diagonal band running from its nose through its eye and a wide "tuning fork "or 'Y' shaped pattern on its forehead.

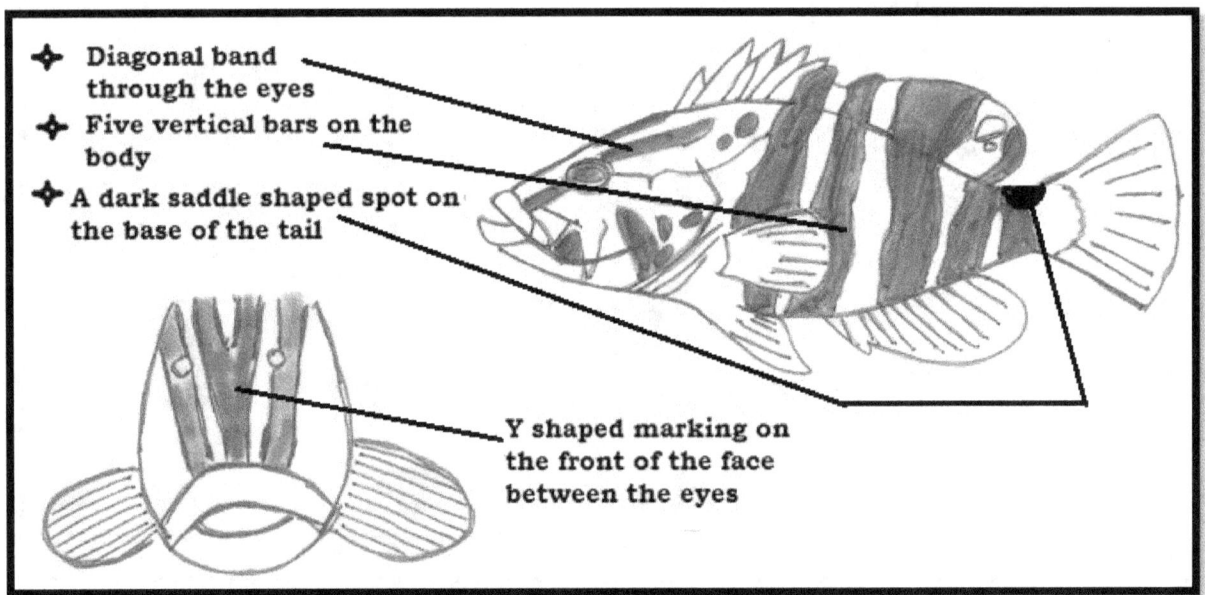

The pictures show distinguishing marks of the Nassau grouper

Reproduction and Life Cycle

The Nassau grouper forms large spawning aggregations made up of few dozen to over 100,000 individuals. They form in depth of 65-130 ft. (20-40 m) during the **full moon of the winter months** – which is late November through February when water temperatures are cool. During this time, most fish have a bi-colored pattern and swim near the bottom. Some females remain in the barred color pattern and become very dark as mating time draws near. A group of bi-colored males swims in circles near the female upon sunset.

A bicolor pattern displayed by both female and male fish is thought to be submissive. A dark phase is displayed by female Nassau grouper that are followed by bicolor fish in courtship. These dark phase females take the lead in spawning events. When eggs and sperm cells are released, it can be seen like a cloud, which is referred to as a sperm cloud.

The eggs are released by the females to be followed by the release of sperm cells by all the bi-colored males. Some eggs are released by some bi-colored females also. This is known as the "spawning rush". Fertilization occurs by chance in the open waters. The larger the spawning aggregations, the better the chances of fertilization occurring.

Life Cycle of the Nassau Grouper

The eggs hatch into pelagic (ocean) larvae that drift along with the currents for a month or so, before becoming juveniles. The larvae have kite-shaped bodies and elongated second dorsal spines. When juveniles are at 10 - 12 months, they float on currents, eventually settling in creeks, the mangroves and inshore algal beds. At about 2" they move to shallow patch reefs. Adult groupers move to deeper reefs where they reach sexual maturity at 4 – 8 years. About one million eggs are released by each female and less than 1% survives to adulthood because they are eaten by other reef fish.

Stages of and Habitats in the Life Cycle of the Nassau Grouper

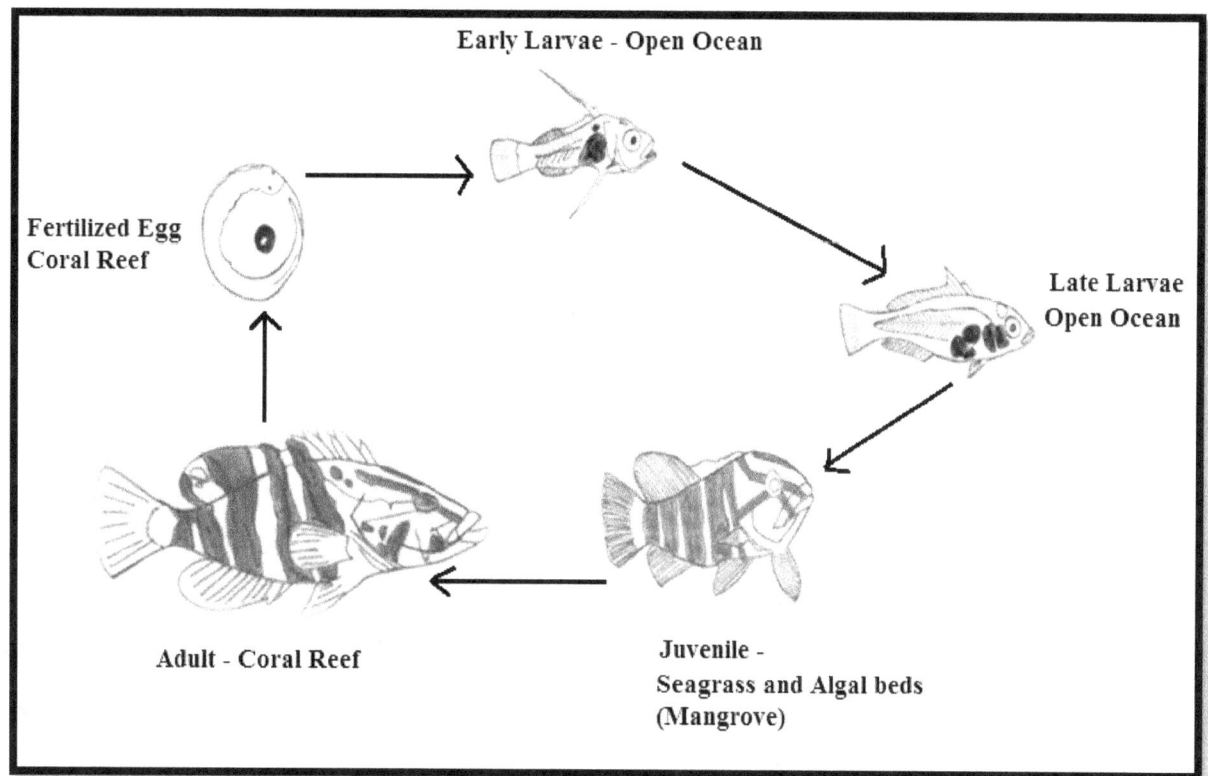

Habitat

Nassau grouper can be found in The Bahamas and South Florida to Central America and northern South America. It can also be found in the Western Atlantic Ocean and the Caribbean Sea to Bermuda.

This grouper is common on offshore rocky bottoms and coral reefs throughout the Caribbean region. They occur at a depth range extending to at least 295 feet (90 m). They rest near or close to the bottom.

Juveniles are found closer to the shore in sea grass beds that offer a suitable nursery habitat. Nassau groupers live apart from other fish but will occasionally form schools. They are active during the day and camouflage themselves to hide from predators. Groupers regularly visit wrasse cleaning stations. At these stations, cleaner wrasses pick parasites and dead tissues from the grouper's gills and body. The grouper will open its mouth to allow the "cleaner fish" to enter its mouth to remove parasites.

Comprehension Questions

1. Draw and label the parts of the Nassau grouper.
2. Give the scientific name for the Nassau Grouper
3. What is the function of the lateral line on the Nassau grouper?
4. State the part of the grouper that is used for smelling.
5. Explain the function of the operculum.
6. The small, thin plates that provide protection for the body and also allow flexibility is known as_____.
7. Name two (2) ways to distinguish the Nassau Grouper from other groupers.
8. Which countries and oceans would you find the Nassau Grouper?

Diet

Nassau groupers are carnivores that ambush their predators. They camouflage themselves and lie in wait among the reef, near tunnels and caves to capture their prey. When prey species come too close, the grouper pounces from their hiding place and engulfs them. They create a strong sucking force, by opening their mouths rapidly and pulling the prey into their jaws. The grouper feeds on shrimp, amphipod, spider crab, spiny lobster, octopus, parrot fish, red snapper and other reef fish. The sucking force of the grouper also helps them in pulling prey out of reef cracks and crevices. Sometimes Nassau groupers drive bottom dwellers out of hiding by stirring up the sand with their tail to uncover prey. The diagram shows a food web showing the Nassau grouper.

Food Chain Showing the Nassau Grouper

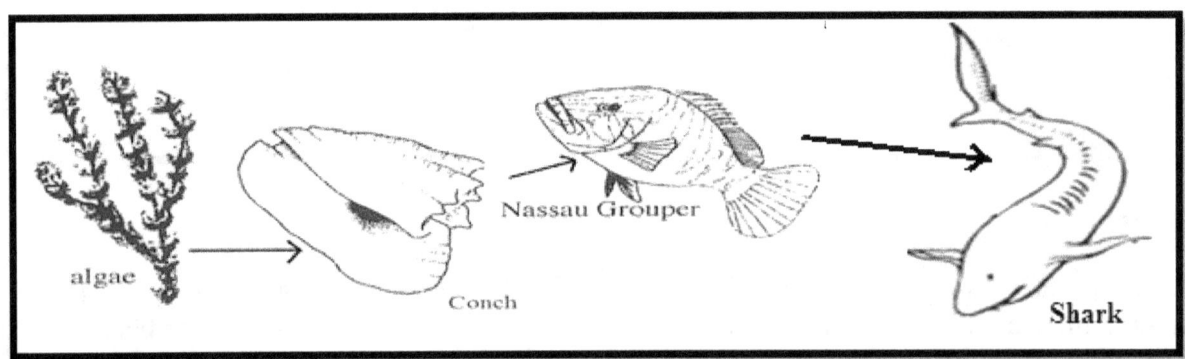

The diagram shows a food web showing the Nassau grouper

Natural Predators

Even though Nassau groupers are one of the reef's largest predators, they are still a prey for larger groupers, sharks and of course, man. Small groupers are eaten by barracudas, lizard fish and dolphins. When predators are nearby, Nassau groupers settle into the coral reef and rapidly pale or darken to camouflage themselves. When they are faced by a predator, the Nassau groupers can make a number of loud booming sounds to discourage them.

Value

- Nassau groupers are a food source for sharks and small groupers are eaten by barracudas, lizard fish and dolphins.
- Some popular dishes made from the grouper are: boiled fish and grouper fingers.
- The Nassau grouper contributes to the local economy.
- The Nassau grouper is also a favourite of snorkelers for its size, inquisitiveness and photogenic appearance.

Effects of Climate Change on Reefs and by Extension, Grouper Populations

Most reefs require a specific water temperature range of 23 to 29 °C for best growth. Some can tolerate higher temperatures, but only for short periods of time. When temperatures fall outside this preferred range, corals can "bleach," when they lose their zooxanthellae and begin to effectively starve. If the temperatures are too high or continue long enough, it results in mass coral mortality. In addition, specific levels of salinity (32 to 42 parts per thousand), water clarity and light levels generally must be consistent throughout the year for corals to grow best. Impacts associated with global climate change, such as increased concentrations of carbon dioxide and other greenhouse gases, can disrupt the delicate balance of the ocean's chemistry. This is called "ocean acidification." Global Warming can elevate seawater temperatures and levels as well, rendering conditions unfit for coral survival. Without the corals, the entire coral-reef ecosystem will be in trouble.

Nassau groupers live on the coral reefs and depend on other marine fish for food and the reef for shelter / survival. Without the coral reef, many reef fish will die or move to other locations for survival. There will be a decrease in the populations of the Nassau grouper and possible extinction of the Nassau grouper if conservation methods are not put in place.

Threats to the Nassau Grouper

- Predation during the closed season.
- Maturing slowly.
- Spawning aggregations are predictable and many fish can be caught before they reproduce.
- Illegal methods of fishing e.g. spear fishing and the capture of juveniles in small mesh traps.
- Habitat destruction - Coral breakage by divers.

- Dredging for construction and siltation from construction runoff.
- Sewage and other contaminants.

Conservation

- The annual closed season from the beginning of December to the end of February.
- Establishment of Marine Parks and Reserves
- Enforcement of the minimum legal harvest size of three (3) pounds
- Discontinue illegal methods of fishing

If the Nassau grouper continues to be overharvested the numbers will continue to decline as the rate of predation exceeds the rate of reproduction. Should there be no restrictions on catching the Nassau grouper, the populations in 25 years will decrease due to overfishing because the grouper will be caught faster than it can reproduce to sustain the population.

The demand for grouper meals will result in an increase in the rate of catching groupers. The higher the demand for the Nassau grouper, the more expensive it becomes.

Increasing the food for Nassau grouper will result in less competition among other predators for food and a better chance of survival.

Measuring the Nassau Grouper

The image below shows the most commonly used measurements for fish.

Total Length Measurement

The **total length** is the maximum length of the fish, with the mouth closed and the tail fin pinched together. The "**fork length**", is measured from the mouth to the center of the opened tail fin. Use the guidelines below to measure the Nassau grouper correctly (total length):

1. Place the fish on its side with the jaw closed.
2. Squeeze the tail fin together or turn it in a way to obtain the maximum overall length.
3. Measure a straight line from the tip of the snout to the extreme tip of the tail fin.

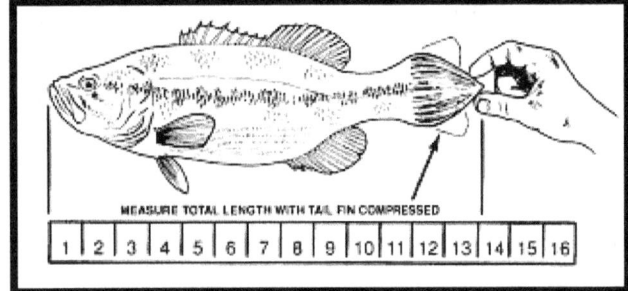

The total length

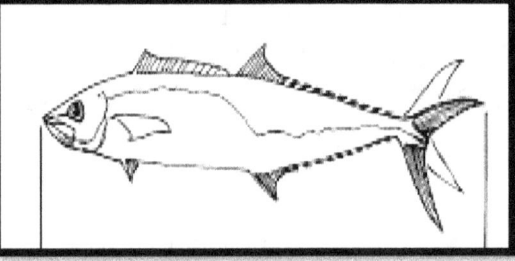

The fork length

Comprehension Questions

1. How does the Nassau grouper hide itself from predators?
2. When preying on species that are in close contact, how does the Nassau grouper use its mouth to catch and eat its prey?
3. What percentages of eggs released by females survive long enough to mature? What are the factors that cause the eggs not to survive?
4. Define the term "spawning rush".
5. Explain the life cycle stages of the Nassau grouper and state where they each settle.
6. Name FOUR (4) threats to the Nassau Grouper.
7. Give ONE (1) way that you can help with the conservation of the Nassau grouper.
8. Explain how you measure the "fork length" and the "standard length" of the Nassau grouper.
9. Draw a food web to include the Nassau grouper.

Activities

- Measure the length of a model grouper in cm and compare the length with the maximum length recorded.
- Use measurements to draw a grouper 10 % of its size.
- Use a balance to measure the weight of a model grouper in kilograms.
- Make a model of the life history of the Nassau grouper.
- Make an oral presentation (with visual aids) to describe the life history of the Nassau grouper.
- Design and produce an infomercial to encourage students to become stewards of the grouper population.
- Produce an item (news flash, comic strip, flyer, fishing item, etc.) to persuade fishermen to observe the restriction laws for fishing grouper.
- Write a letter to the Minister of Agriculture and Marine Resources advocating for a particular law, method of catching or boosting grouper populations.
- Present a case for the use of alternative fish (e.g. "rock fish") in the grouper family for use instead of the Nassau grouper.
- Participate in a debate on the importance of observing the closed season for catching the grouper OR the economic importance of the grouper outweighs the need to protect it.

Particles of an Atom

Inside of an atom, there are three subatomic particles: **protons**, **neutrons**, and **electrons** (as seen in the Helium atom below). A **subatomic particle** is a particle smaller than an atom. This means it is very, very small. Like atoms and molecules, a subatomic particle is far too small to be seen with the naked eye.

Most of an atom's mass is in the *nucleus* which is found at the center of every atom. In the nucleus are protons and neutrons. Protons have a positive (+) charge. Neutrons are neutral, meaning they have no (0) charge. Electrons have a very small mass and are negatively (-) charged. The mass of an electron is equal to 1 / 1836 of a proton. Electrons are located outside of the nucleus and move around the atom's nucleus.

Structure of an Atom

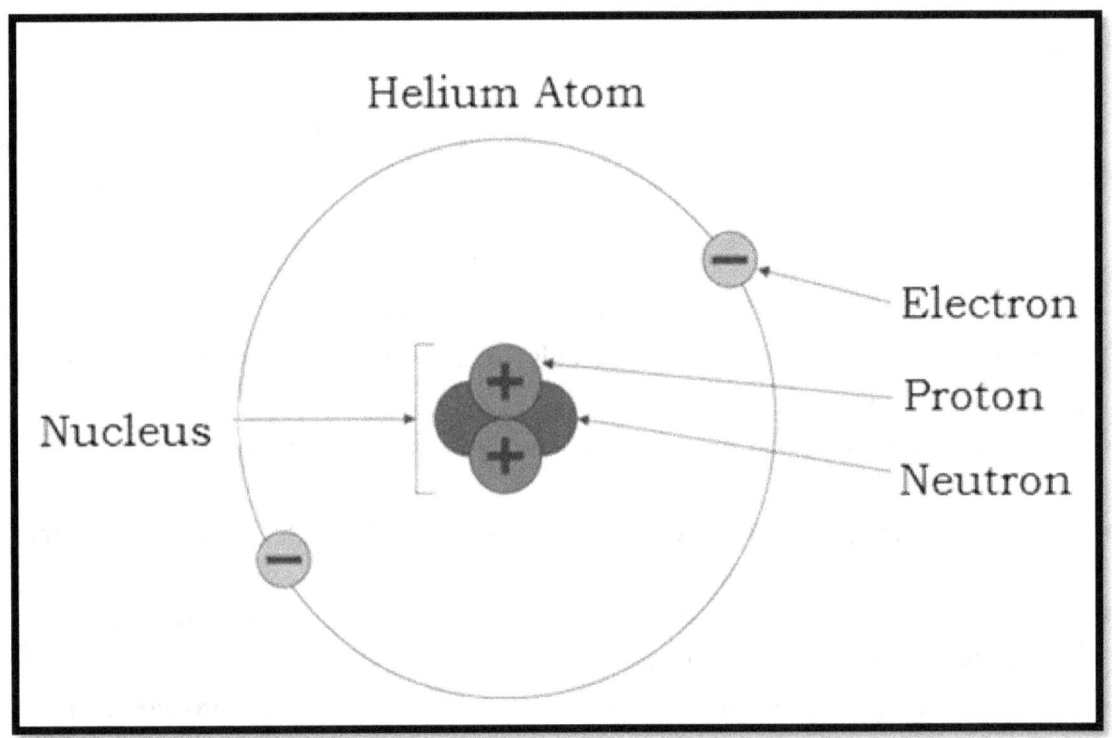

Comparison of Subatomic Particles

Subatomic Particle	MASS	CHARGE
Proton (found in nucleus)	1	Positive (+)
Neutron (found in nucleus)	1	Neutral (0 / No)
Electron (orbits around nucleus)	1 / 1836 of a proton	Negative (-)

Electronic Configuration of an Atom (Oxygen)

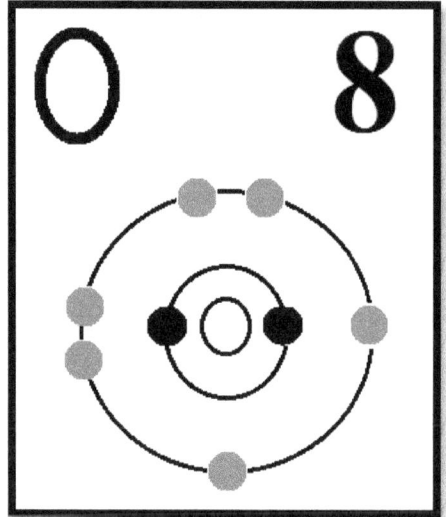

Electronic configuration of an oxygen atom

The distribution of electrons in the various shells (orbits) is called electronic arrangement or **electronic configuration**. Electronic arrangement can be shown by numbers or you can show it by drawing the atom, For example, you can draw the atom as shown above or you can write the numbers, O (2, 6).

The oxygen atom shown in the above diagram has 8 protons, 8 neutrons in the nucleus and 8 electrons in its orbit. The number of electrons is the same as the number of protons. The electronic configuration has 2 electrons in the inner orbit and 6 electrons in the second orbit. You can use the following symbols to show the electrons: ●, ✗ or ◯

Each electron shell is given a number 1, 2, 3, or 4. Each level can hold up to a certain number of electrons. For example, Level 1 can hold a maximum of 2 electrons. Level 2 can hold a maximum of 8 electrons. Level 3 can hold a maximum of 18 electrons and level 4 can hold a maximum of 32 electrons but, the **outermost level** of any atom can hold no more than 8 electrons.

Boron has an atomic number of 5. This means that there are 5 electrons around the nucleus. The first energy level can only hold 2 electrons, so the remaining 3 must go to energy level 2, it can be written, B 2, 3.

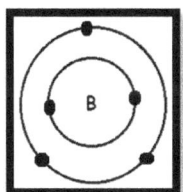
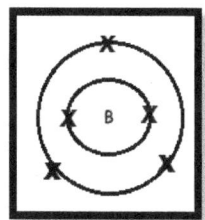

Electronic Configuration of a Boron Atom

A chemical element is identified by the number of protons in its nucleus and it must collect an equal number of electrons if it is to be electrically neutral. The first shell can have only 2 electrons, so that shell is filled in helium on the periodic table; it is the first noble gas. In the periodic table, the elements are placed in "periods" and arranged left to right in the order of filling of electrons in the outer shell. So hydrogen and helium complete the first period.

Chemical Reactions

Electrons in their shells around an atom are the basis of chemical reactions. Complete outer shells, with the maximum number of electrons, are less reactive. Outer shells with less than maximum electrons are reactive. The number of electrons in atoms is the underlying basis of the chemical periodic table.

Comprehension Questions

1. Name the 3 subatomic particles within an atom.
2. Give the charge for each subatomic particle.
3. Study the diagram of the structure of an atom and tell the location of the nucleus.
4. Where is most of the mass of an atom found?
5. Which subatomic particle orbits the nucleus?
6. Draw the structure of the Helium atom and identify the parts.
7. Define the term 'Electronic Configuration'.
8. A _____ _____ is identified by the number of protons in its nucleus.
9. Compare the relative mass of electrons and protons by making 1836 dots in pencil to one dot in pen.

Electronic Arrangements

Give the full electronic configuration in the following. The first three has been done for you.

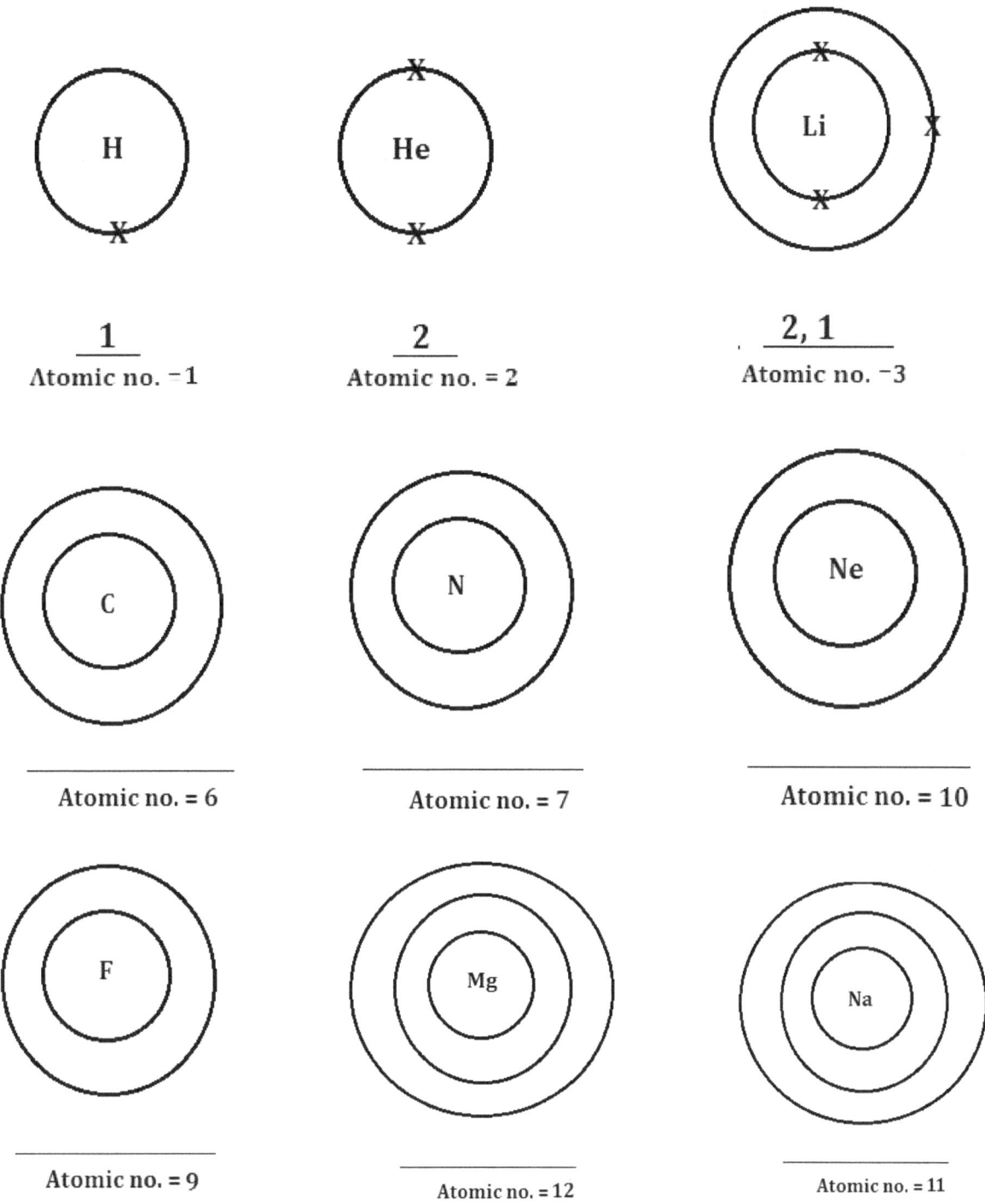

1

Atomic no. = 1

2

Atomic no. = 2

2, 1

Atomic no. = 3

_____

Atomic no. = 6

_____

Atomic no. = 7

_____

Atomic no. = 10

_____

Atomic no. = 9

_____

Atomic no. = 12

_____

Atomic no. = 11

Elements and the Periodic Table

Elements are pure substances made of only one kind of atom. Atoms are tiny structures found in all matter. Elements cannot be broken down further into other substances. Most substances contain many different atoms; however, elements contain only ONE kind of atom. Elements are the building blocks of matter. There are over one hundred elements. Elements are named after people, places and things.

The periodic table is a means of classifying elements according to their characteristics. It is a chart that shows information about all of the elements, including their chemical symbols. At first scientist used a single capital letter for chemical symbol. For instance, H stands for hydrogen and O stands for oxygen. Later, scientist created two-letter abbreviations for elements. The chemical symbol for aluminum is Al and the chemical symbol for magnesium is Mg. Some elements were given symbols based on their Latin names; therefore, their symbols are very different from their English names. The chemical symbol for iron is Fe because the Latin word for iron is Ferrum. A symbol must begin with a capital letter and all other letters should be in lower case.

Below is a list of the first twenty elements and symbols in the periodic table.

Hydrogen	H	**Sodium**	Na
Helium	He	**Magnesium**	Mg
Lithium	Li	**Aluminum**	Al
Beryllium	Be	**Silicon**	Si
Boron	B	**Phosphorus**	P
Carbon	C	**Sulphur**	S
Nitrogen	N	**Chlorine**	Cl
Oxygen	O	**Argon**	Ar
Fluorine	F	**Potassium**	K
Neon	Ne	**Calcium**	Ca

The Periodic Table of Elements

Understanding the Periodic Table

The Periodic Table is a list of all the elements that are known. It is organized by increasing atomic number. There are two main groups on the periodic table: **metals and nonmetals**. Most elements are metals. The left side of the table contains elements with the greatest metallic properties. As you move from the left to the right, the elements become less metallic with the far right side of the table consisting of nonmetals. The elements in the middle of the table are called "**transition**" elements because they are changed from metallic properties to nonmetallic properties. A small group, whose members touch the zigzag line, are called **metalloids** because they have both metallic and nonmetallic properties.

The table is also arranged in **vertical columns** (going up and down) called "**groups" or "families"** and **horizontal rows** (going across) called "**periods.**" Each arrangement is important. **The elements in each vertical column or group have similar properties**. Group 1 elements all have the same number of electrons in their outer shells. This gives them similar properties. Group 2 elements all have 2 electrons in their outer shells. This also gives them similar properties. From the top to the bottom of a group, trend shows elements are less reactive (electrons in outermost orbit are further from the nucleus).

Elements in a period have the same outermost orbit (K, L, M and N). The elements in the first period or row all have one shell. The elements in period 2 all have 2 shells. The elements in period 3 have 3 shells and so on. Successive elements across a period show an increase in the number of electrons in the outermost orbit.

Elements in Groups 1, 2, 6 and 7 as well as elements in Periods 1, 2 and 3 tend to be more reactive.

There are a number of major groups with similar properties. They are as follows:

Hydrogen: does not match the properties of any other group so it stands alone. It is placed above group 1 but it is not part of that group. It is a very reactive, colorless, odorless gas at room temperature. (1 outer level electron)

Group 1: Alkali Metals – are enormously reactive and are never found in nature in their pure form. They are silver colored and shiny. Their density is extremely low so that they are soft enough to be cut with a knife. (1 outer level electron)

Group 2: Alkaline-earth Metals – are slightly less reactive than alkali metals. They are silver colored and denser than alkali metals. (2 outer level electrons)

Groups 3 – 12: Transition Metals – have a moderate range of reactivity and a wide range of properties. They are shiny and good conductors of heat and electricity. They also have higher densities and melting points than groups 1 & 2. (1 or 2 outer level electrons)

Lanthanides and Actinides: are also transition metals that were taken out and placed at the bottom of the table so the table would not be so wide. The elements in each of these two periods share many properties. The lanthanides are shiny and reactive. The actinides are all radioactive and are unstable. Elements 95 through 103 are not natural but have been made in the lab.

Group 13 or (Group 3): Boron Group – contains one metalloid and 4 metals. They are reactive. Aluminum is in this group. It is also the most abundant metal in the Earth's crust. (3 outer level electrons)

Group 14 or (Group 4): Carbon Group – contains one nonmetal, two metalloids, and two metals. There is varied reactivity in this group.   (4 outer level electrons)

Group 15 or (Group 5): Nitrogen Group – contains two nonmetals, two metalloids, and one metal. There is varied reactivity in this group.   (5 outer level electrons)

Group 16 or (Group 6): Oxygen Group – contains three nonmetals, one metalloid, and one metal. They are reactive.   (6 outer level electrons)

Groups 17 or (Group 7): Halogens – contains all nonmetals. They are very reactive and poor conductors of heat and electricity. They tend to form salts with metals. E.g. NaCl: sodium chloride also known as "table salt".   (7 outer level electrons)

Groups 18 or (Group 8): Noble or inert Gases – are un-reactive nonmetals because they have the required number of electrons in the outermost orbit to keep them stable. They are colorless, odorless gases at room temperature. All of them are found in the Earth's atmosphere in small amounts.   (8 outer level electrons, helium has 2 electrons).

Physical versus Chemical Properties

A physical property is observed with the senses and can be determined without destroying the object. For example, color, shape, mass, length, odour, solubility, density, melting point, boiling point, strength, hardness, ability to conduct electricity and magnetism are all examples of physical properties.

A chemical property indicates how a substance reacts with something else (chemical reaction). When a chemical property is observed, the original substance is changed into a different substance. The ability of iron to rust is a chemical property. The iron reacts with oxygen and the original iron metal is gone. It is now iron oxide, a new substance.   Other examples; Wood burns in air. Sodium burns in chlorine gas. Magnesium reacts with vinegar to form hydrogen gas. All chemical changes include physical changes.

Differences between Metals and Non-metals

METALS	NON-METALS
1. Shiny	Dull
2. Solid except mercury	Can be liquids, gases or solids
3. Good conductors of heat	Poor conductors of heat
4. Good conductors of electricity	Poor conductors of electricity
5. Make a noise when hit.	Little or no noise when hit
6. Poor insulator	Good insulator
7. Malleable and ductile (can be hammered into shape and drawn into wires).	Brittle when solid
8. High melting and boiling points	Have relatively low melting and boiling points
9. Usually dense	Much less dense
10. React with dilute acids and give off hydrogen gas.	Does not react with dilute acid.

Chemical Properties of Metals and Non- Metals

Metals	Non-Metals
Usually have 1-3 electrons in their outer shell	Usually have 4-8 electrons in their outer shell.
Lose their electrons easily.	Gain or share electrons easily.

Comprehension Questions

Periodic Table Tasks

1. Use the Periodic table above and divide the metallic and non-metallic elements by a thick zigzag line.

2. Which type of element are most elements?
3. List four typical physical properties of (a) metals, (b) non-metals.
4. List two characteristic chemical properties of metals.
5. What is meant by:

    a. Group of elements? Why is it so important that they are 'grouped' in this way?
    b. Period of elements? - How does the general character of the element change from left to right?
6. Mark the Group numbers 1 - 8 at the top of the correct column.
7. Down the side mark the Period number from 1, down to 6.
8. Name the first and last element in periods 1 to 3.
9. What is the connection between an element's electron arrangement and its Group number?
10. What is the connection between an element's electron arrangement and its Period number?
11. Draw the following electron structures:

(A) 2, 8, 7 (B) 2, 4, (C) 2,8,8,1

12. List three noble gases.

Comprehension Questions

Follow the instructions below to label the major groups and divisions of the periodic table.

1. The vertical columns on the periodic table are called _____.

2. The horizontal rows on the periodic table are called _____.

3. Most of the elements in the periodic table are classified as _____.

4. The elements that touch the zigzag line are classified as _____.

5. The elements in the far upper right corner are classified as_____.

6. Elements in the first group have one outer shell electron and are extremely reactive. They are called _____ _____.

7. Elements in the second group have 2 outer shell electrons and are also very reactive. They are called _____ _____ _____.

8. Elements in groups 3 through 12 have many useful properties and are called _____ _____.

9. Elements in group 17 are known as "salt formers". They are called _____.

10. Elements in group 18 are very un-reactive. They are said to be "inert". We call these the _____ _____.

Mixtures and Solutions

A **mixture** is two or more substances combined together that may be easily separated physically. A mixture can contain elements, compounds, or both in any amounts. In a mixture, each substance keeps its own properties.

There are different kinds of mixtures. A **homogeneous mixture** has particles spread evenly throughout the substance. Tea, root beer, and vinegar are examples of homogeneous mixtures. Each part of the mixture is exactly like every other part.

A **heterogeneous mixture** is a mixture in which the particles are not spread evenly throughout. Soil, raisin bran cereal, and fruit salad are heterogeneous mixtures.

Examples of homogeneous and heterogeneous mixtures are shown below.

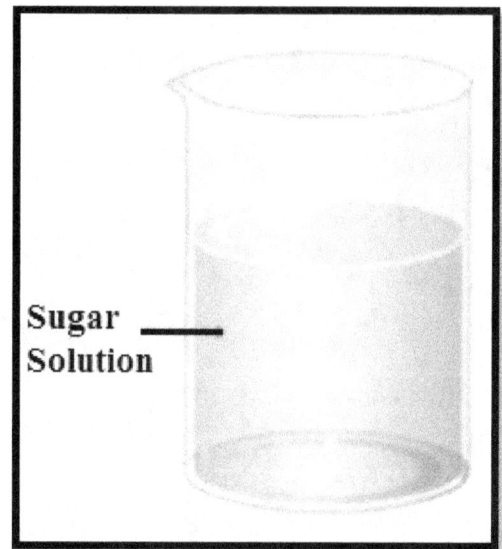

Sugar Solution

A Homogeneous Mixture

A Heterogeneous Mixture – Fruit Salad

A **solution** is two or more substances combined together that may not be separated physically within a solution, one substance is dissolved (completely disappears) in another substance. A solution is a homogeneous mixture. The substance that dissolves is called a **solute**. The substance into which a solute dissolves is called a **solvent**. Water is a common solvent and is called the **universal solvent** because many substances can be dissolved in water. Other solvents are alcohol, turpentine and vinegar.

An **aqueous solution** is a **solution** in which the **solvent** is water.

Solubility is the ability of a substance to dissolve in another substance. Sugar is a soluble substance that dissolves easily in water. Carbon dioxide is soluble. It is dissolved in soda to make it "fizz."

A **suspension** is when very fine particles of a solid are mixed with a liquid. The solid is in suspension so the water will look cloudy. The particles in a suspension are not dissolved; they will settle out from the force of gravity. Shaking or stirring will suspend the particles again. Many liquid antibiotic prescribed by Pediatricians are suspensions. This is why you have to shake the contents of the bottle before use. Another example is muddy water.

A Suspension

A

B

A. Muddy water

B. Mud settles at the bottom and the water become clear at the top

A Colloidal Mixture

A colloidal mixture is a substance which is suspended throughout the solvent to give it a homogeneous appearance. The particles in a colloidal mixture are of a size between that of a solution and a suspension. The particles **do not settle** out with gravity and cannot be filtered out. Examples of colloidal mixtures: milk and Pepto Bismol are shown below.

Pepto-Bismol

Milk

An **unsaturated** solution is one in which more of the solute could dissolve in the solvent at the same temperature.

A **saturated** solution is one in which *no* more of the solute will dissolve in the solvent at a specific temperature.

A **supersaturated** solution is when a solution contains more solute than would normally dissolve at a certain temperature. When you heat the solvent more of the solute can be dissolved. Generally, crystals dissolve in water. An example of this is the syrup that is used in snow cones. It is made from a lot of sugar dissolved in hot water.

Snow Cone made with a Supersaturated Solution

A supersaturated solution- Syrup

Syrup used on snow cone

Dilute and Concentrated Solutions

A **concentrated solution** is a strong solution with more of the solute dissolved in the solvent. A concentrated solution is a strong solution. The amount of solute in a solvent is a measure of the concentration of the solution. The larger the amount of solute for a given volume of solvent, the greater the concentration of the solution, termed concentrated solution

A **dilute solution** has relatively little solute. It is a weak solution.
Increasing the amount of water in a solution of water and a known solute will eventually cause the solution to become diluted.

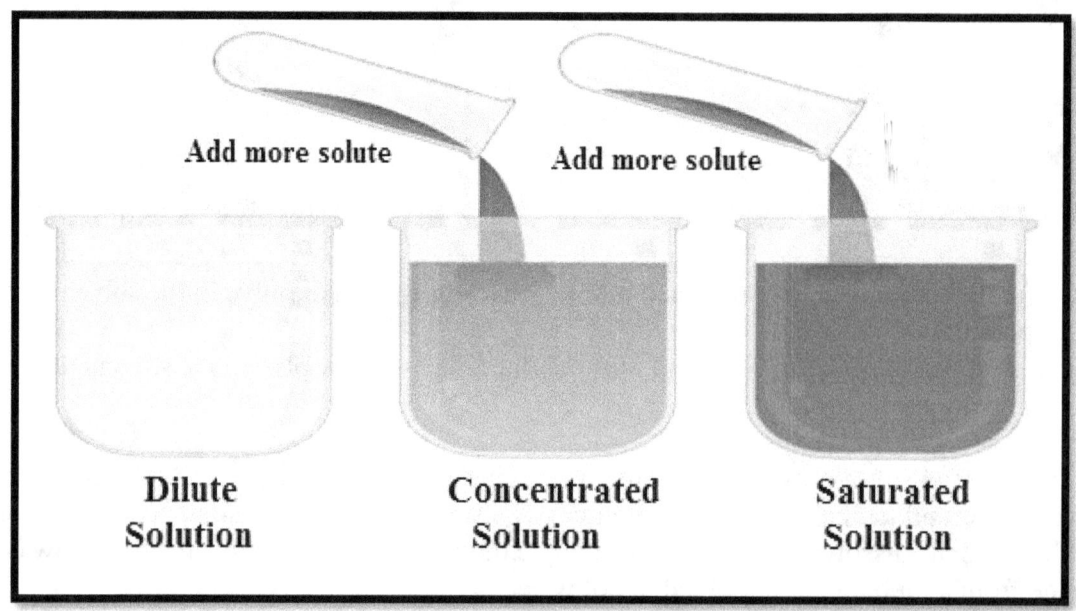

The pictures show dilute, concentrated and saturated copper sulphate solutions

Factors Which Affect the Solubility of a Solute

As **temperature** is increased, solubility of a substance increases and more of the solute will dissolve in the solvent.
As **pressure** increases the solubility of a gas increases.
As the **amount of solvent** increases, the solubility of a substance increases.

Control Variables When Comparing Solubility of Substances

A variable is any factor that can be changed or controlled in an experiment. Temperature, pressure, constant solvent, volume of solvent and the nature (form/appearance) are all variables that can change if they are not controlled. Avoid any conditions that increase or decrease the solubility of a substance contrary to its intended state.

The mass of the solute and the volume of the solvent used in an activity will affect the solubility of the solute in the solvent. Note: Mass is measured in grams to (0.1g accuracy).

The larger the amount of solute dissolved in a solvent at a given temperature, the less soluble the solute becomes in that solvent.

Separation of Mixtures

The different substances in mixtures are usually easily separated from one another because they are not combined chemically. The method you use in separating mixtures depends upon the type of mixture. .

Filtration

Filtration is good for separating insoluble particles from a liquid. An insoluble particle is one that does not dissolve. Sand, for example, can be separated from a mixture of sand and water using filtration. That is because sand does not dissolve in water. The filter paper that is used in the separation has tiny pores and acts like a sieve allowing the water (filtrate) to pass through and holding back sand which are the larger particles (residue). When pouring the mixture into the filter paper, the mixture must be kept below the top of the filter paper to avoid contamination of the filtrate. All particles or molecules smaller than the pore size of the paper can pass through (like salts and microbes).

The solution containing the suspended impurities is made to pass through the porous membrane of the filter paper, filter cloth, etc. The water or solvent containing any dissolved substances passes through the porous membrane, which is called filtrate. The solid suspended particles that remain on the porous membrane (filter paper) is termed residue.

Filtration

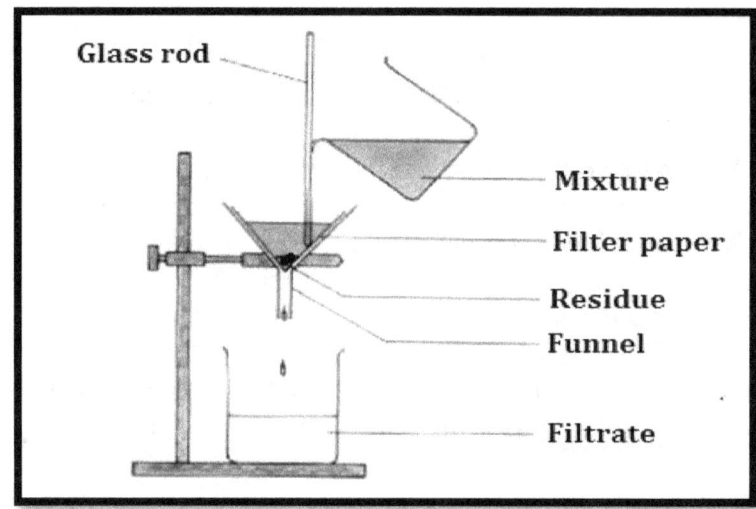

Separation of a Mixture by Filtration

Separation of a Mixture by Decantation or Sedimentation

When separating a mixture of coarse particles of a solid from a liquid (muddy water) by decantation, the mixture is taken in a container and allowed to stand for some time. The coarse particles of the solid being heavier than the liquid (usually water), settle down due to gravity. Bigger particles settle down faster than the finer particles. Settling down of the particles leaves the upper layers of the liquid clearer. The clear upper layer of the liquid is then gently poured out into another container. Settling down of the coarse particles due to the effect of gravity is called **sedimentation.** The transfer of the clear upper liquid without disturbing the settled solid particles is called **decantation**.

Sedimentation and Decantation

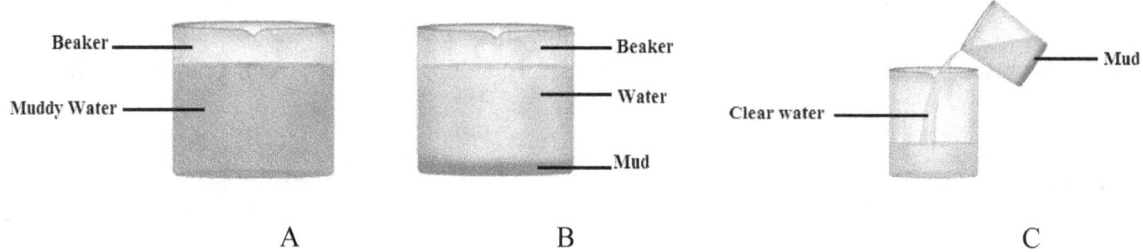

Separation of the coarse particles of a solid from a liquid by sedimentation and decantation

a. Muddy water

b. Sedimentation - Mud settles at the bottom and leaves the upper layer of liquid clearer.

c. Decantation - Water is poured off thereby separating the mixture

Other example of decantation: Oil forms a layer on water; It is possible to carefully pour off the oil or let the water out.

Separation of a mixture by magnetism

A magnet attracts a magnetic material out of a mixture containing it. The magnetic component of the mixture is separated by the magnetic attraction. A magnet is moved over the mixture containing the magnetic substance e.g., iron filings. These get attracted to the magnet. The process is repeated until the magnetic material is completely separated from the mixture.

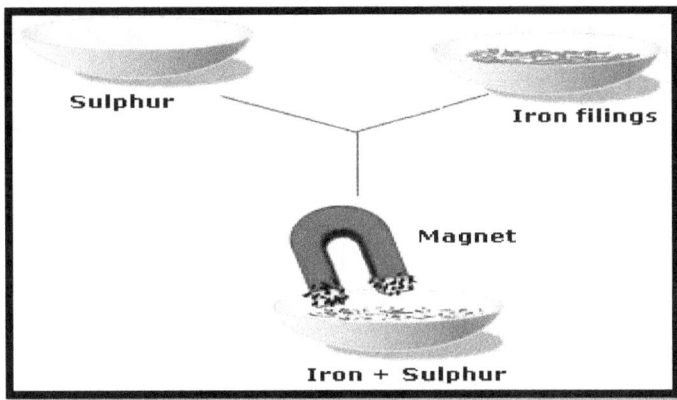

Separation by Magnetism

Evaporation

This is good for separating a **soluble solid** from a liquid (a soluble substance dissolves to form a solution). The solvent will have a lower boiling point than the solute and will boil off leaving the solute. Evaporation of water begins at 100°C. Evaporation becomes faster at higher temperatures. The solution containing the mixture is taken in an evaporating dish and heated gently. Gradually the solvent or water evaporates and the solution containing the dissolved solute becomes thicker and heated until it is dry. When all of the liquid evaporates, the solute will remain behind. In the example below, the solution is salt and water. When the water evaporates, the salt is left behind.

Process of Evaporation

The diagrams show how evaporation takes

Chromatography

Paper chromatography is a technique used to separate a mixture into its component molecules. Chromatography is good for separating dissolved substances that have different colours, such as inks and plant dyes. Dyes differ in their relative weight because of the differences in solubility, mass, and hydrogen bonding with the paper. It therefore moves different distances during separation. The lighter dyes or colours will move furthest away from the original spot.

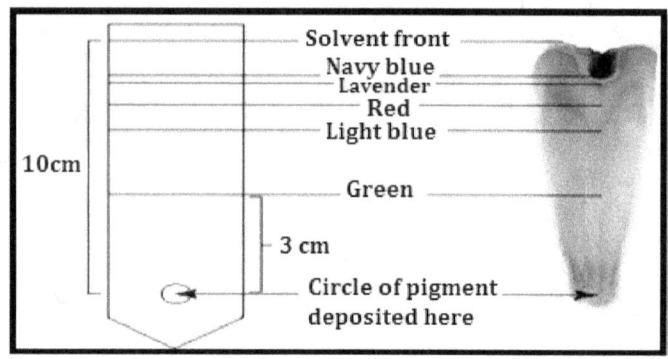

Separation of Ink by Chromatography

Distillation

A solution containing components with different boiling points may be separated into its components. This separation can be done through distillation when each component reaches its boiling point. It evaporates then condenses into a container. Listed below are the steps in the distillation process:

- Place the impure water into a flask and heat the water using a Bunsen burner.
- The impure water will turn into steam and pass through the Liebig's condenser.
- The cold water in the condenser will cool the steam and it will become water droplets.
- The water droplets (water) will then flow and collect into the beaker. This is a very pure form of water called distilled water.

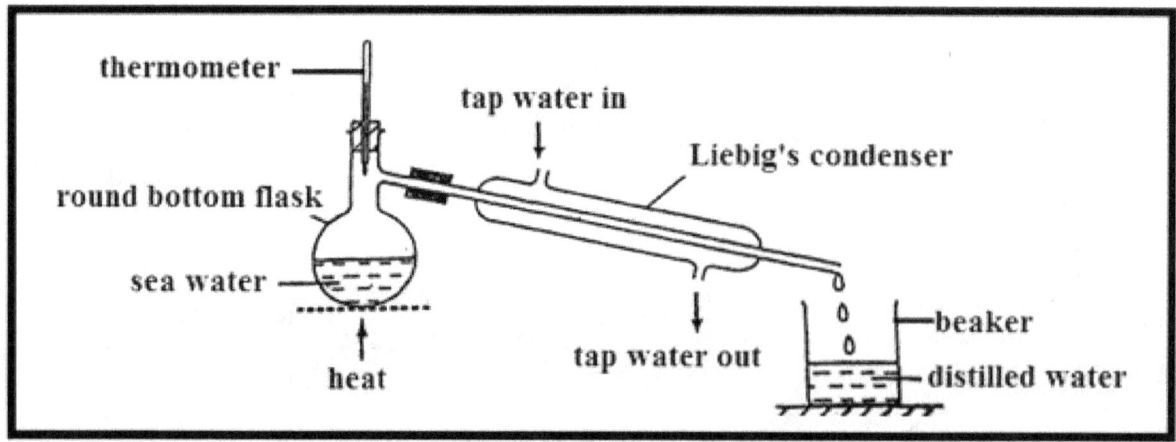

The diagram shows the distillation process

Less conventional methods for separating mixtures in the food preparation and construction industries

Using sieves
Hand sorting
Pouring off less dense substances
Use of a mechanical press (pressing)

Centrifugation
Electromagnetism
Drainage

Methods to Extract Coconut Oils
Oils can be pressed from beans and nuts like coconuts using a mechanical press.

Sublimation of Iodine Crystals

The term **Sublimation** means to change from a solid to gaseous state without a liquid phase because the melting and boiling points are very close. Sublimation happens in Iodine.
There is a larger difference between the melting and boiling points of water and a number of other chemicals; therefore they do not undergo sublimation under laboratory conditions.

Comprehension Questions

1. Define the following terms:
   a. Mixture
   b. Heterogeneous mixture
   c. Homogeneous mixture
   d. Solutions
   e. Solute
   f. Solvent
   g. Universal Indicator
   h. Solubility
   i. Saturated Solution
   j. Unsaturated Solution
   k. Super Saturated Solution
   l. Aqueous Solution
2. Why is water called the universal solvent?
3. List two solvents other than water.
4. What are the differences between concentrated and dilute solutions.
5. What factors affect the solubility of a solute?
6. List FOUR Separation Techniques.
7. Which separating technique is best for separating insoluble solids?
8. With the use of diagrams, explain the Decantation process.
9. Explain why magnetism would be the best method to be used to separate sulphur and iron filings.
10. At what temperature does evaporation takes place?
11. Which separating technique separates dissolved substances that have different colors?
12. Why do dyes differ in their relative weight?
13. What are the Steps of Distillation?

Machines

A machine is a device that makes work easier. Tasks may be very difficult or impossible otherwise. It enables us to use a small amount of force (effort), to move a larger force (load). A machine is sometimes called force multipliers. A simple machine is made up of a few moving parts or none at all. **Simple machines** require **human** energy in order to function. Simple machines DO NOT need motors. They can change the size or direction of a force in one motion. There are six types of "simple" machines. Many of these machines have been used since prehistoric times.

Types of Simple Machines

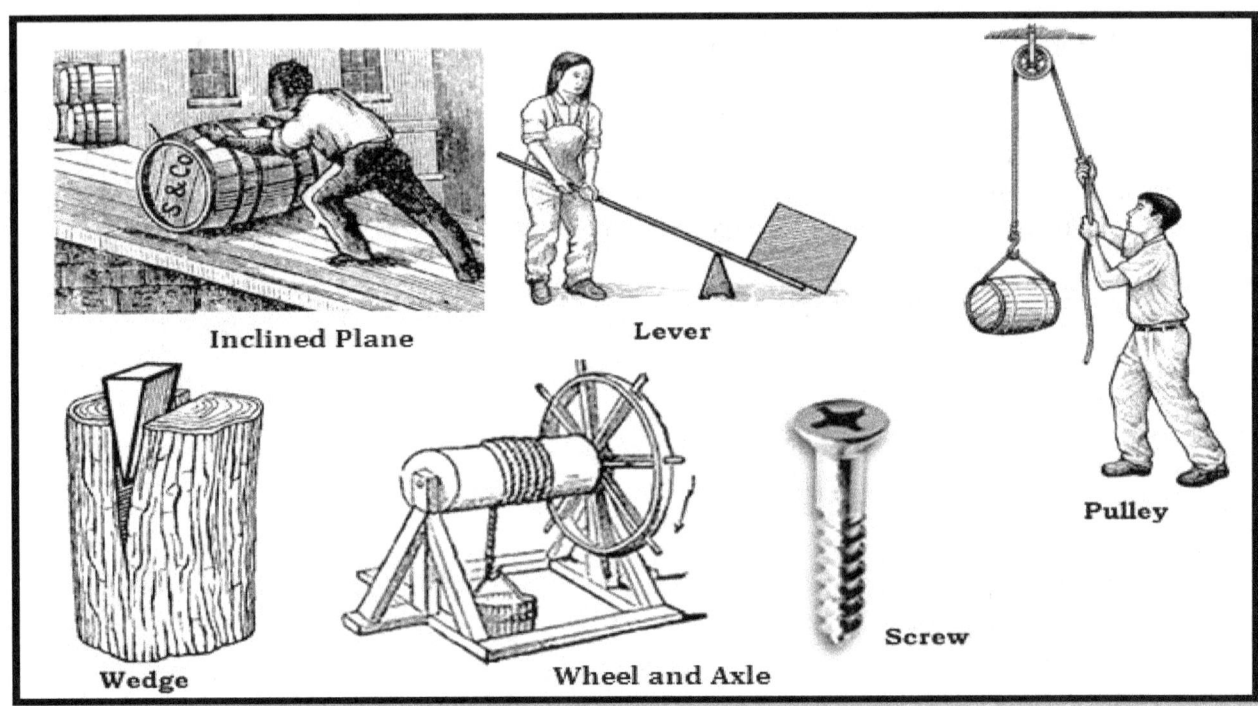

The pictures show the six types of simple machines

All machines are made from variations of simple machines. All the six simple machines can be put together to make big machines. The machine that is made up of two or more simple machines working together is called a **complex machine.** A complex machine sometimes requires the use of a motor. E.g. cars, trucks, power tools, sewing machines, etc. The following is a list of what machines do:

Machines help us do *work more easily*. Machines *do not* save work from being done however.

1. A machine can make *the force* you put into a machine *greater*. (ex. Pliers)
2. A machine can *change the direction* of the force you put in. (ex. A Car jack)
3. A machine can *increase the speed* of the force. (Ex. Bicycle)

Machines have what is called **mechanical energy**, (the energy of moving things). Some examples of mechanical energy are: wind, moving rocks, and waterfalls. The energy output compared to the energy input is called **efficiency**.

1. Levers

The Lever	
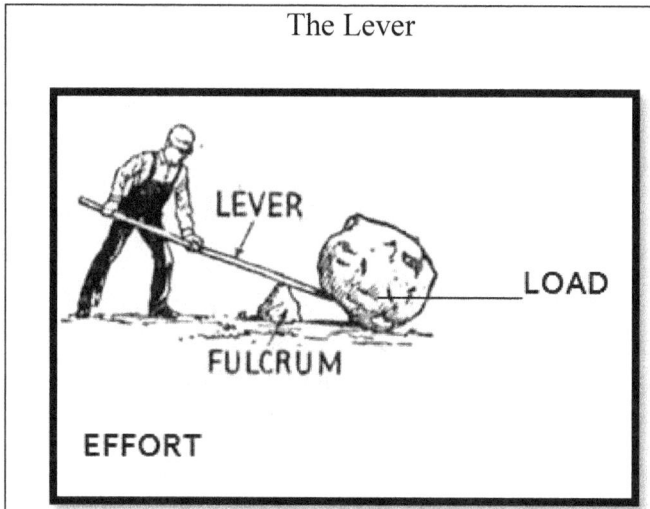	A lever is a rigid support that pivots on a point. In the picture to the left you will find a diagram detailing the pivot point (fulcrum) and the support bar (lever).

Types of Levers

A lever is a simple machine made of a rod or a bar that is used to move objects. It turns or pivots on a fixed point to move the load or weight. All levers have three main parts:

1. The fulcrum is the point on which the lever is supported and turns.
2. The effort is the part on which the force is applied by pushing or pulling.
3. The resistance or load is the part which bears the weight to be raised.

Although all levers have these three parts, the **three parts can be arranged in different places.** Levers come in three classes. Depending on what is in the **middle,** determines what class of lever it is.

First Class Levers (Class 1 Levers)

The **first class lever** has its fulcrum in the middle, between the load and effort.   E **F** L. The closer the fulcrum is to the load, the less force is required to move the load. E.g. a seesaw, claw hammer, crowbar, scissors, pliers, and tin snips.

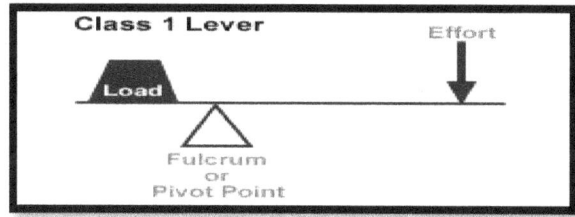

First class lever

First Class Levers

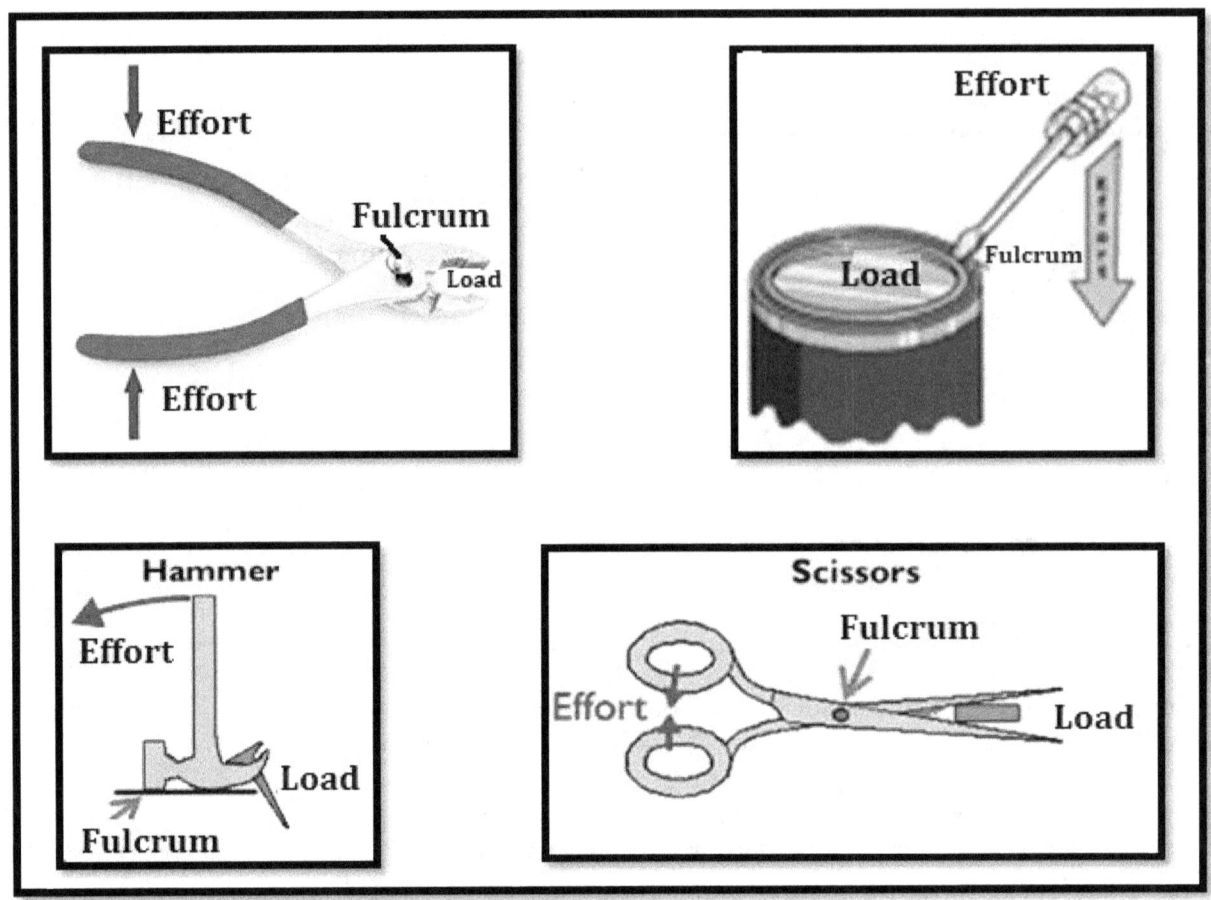

The pictures show class 1 levers

Second Class Levers (Class 2 Levers)

The **second class lever** has its load in the middle, between the effort and the fulcrum.   E **L** F.

E.g. a wheelbarrow, paper cutter, door, nutcracker, garlic press, bellows, and a bottle opener.

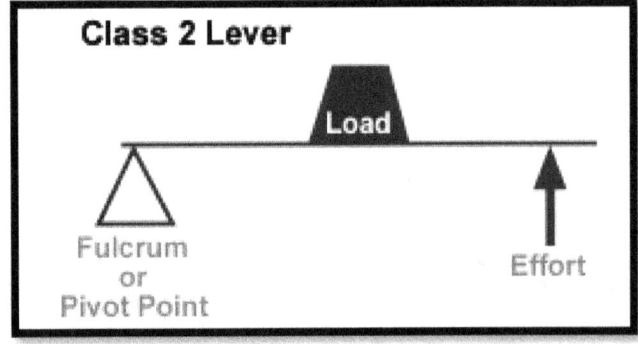

Second class lever

Second Class Levers

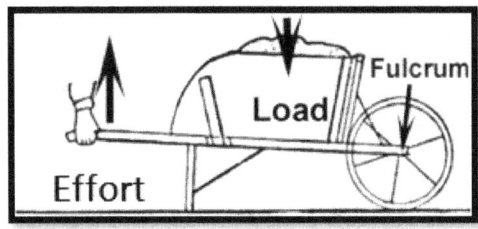

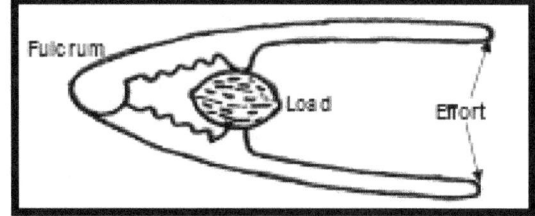

Wheel barrow   Nut cracker

Third Class Levers (Class 3 Levers)

The **third class lever** has its effort in the middle, between the fulcrum and the load.   F _E_ L

E.g. oar, a fishing pole, hammer, baseball bat, hockey stick, golf club, tennis racket, shovel, pitchfork, hoe, broom, tweezers, ice tongs, and your arms and legs.

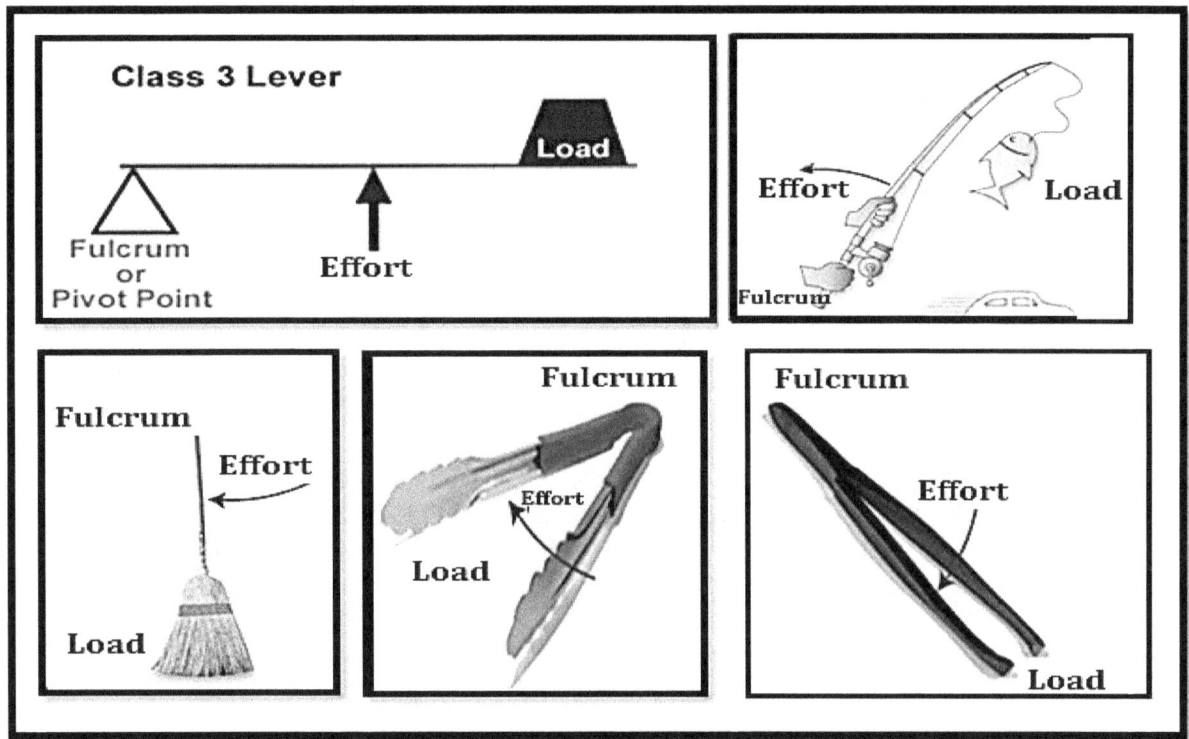

Class 3 levers

Mechanical Advantage

Mechanical advantage is a measure of how much a machine multiplies the effort (distance or force) to make the task easier. The mechanical advantage in a lever can be calculated using different lengths of effort arm and load arm.

Mechanical Advantage of Lever = $\dfrac{\textbf{Length of effort arm}}{\textbf{Lenght of load arm}}$

To calculate the effort required to lift a load using a pulley you divide the load by the number of ropes (do not count the rope that goes to the effort).

2. Pulleys

A fixed pulley has a fixed axle. The axle is "fixed" or anchored in place. A fixed pulley is used to change the direction of the force on a rope (called a belt). A fixed pulley has a mechanical advantage of 1. A mechanical advantage of one means that the force is equal on both sides of the pulley and there is no multiplication of force.

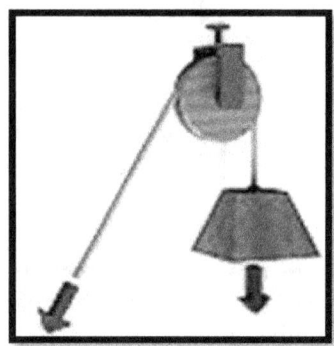

Fixed pulley

A movable pulley has a free axle. The axle is "free" to move in space. A movable pulley is used to multiply forces. A movable pulley has a mechanical advantage of 2. If one end of the rope is anchored, pulling on the other end of the rope will apply a doubled force to the object attached to the pulley.

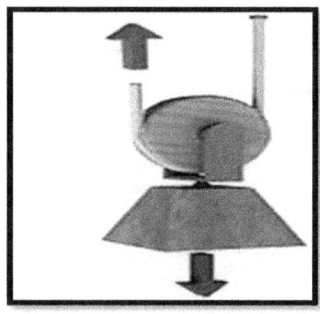

Movable pulley

A Compound Pulley is a combination of a fixed and a movable pulley system.

Block and Tackle - A *block and tackle* is a type of compound pulley where several pulleys are mounted on each axle to increase the mechanical advantage. Block and tackles usually lift objects with a mechanical advantage of greater than 2.

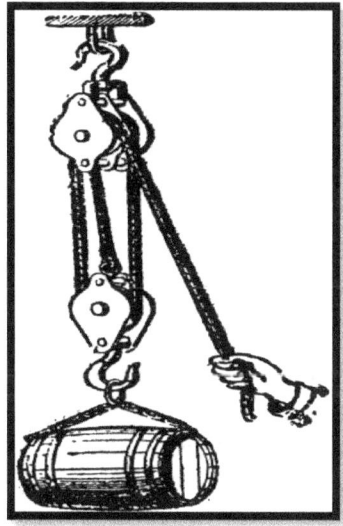

Block and Tackle

3. Inclined Planes are slanted surfaces that help you do work. Some examples are stairs and ramps. Inclined planes along with the other simple machines provide us with an advantage. The formula for the mechanical advantage of an inclined plane is the length divided by the height.

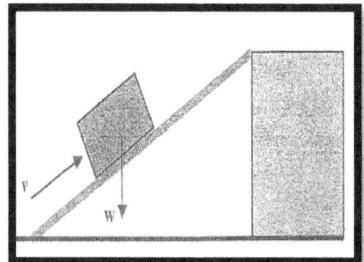  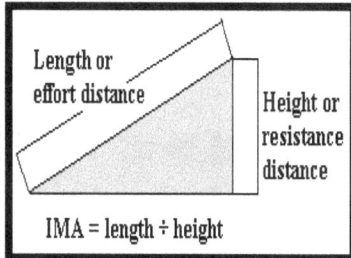

The pictures show examples of inclined planes

4. A wedge is two inclined planes joined together usually used for cutting. An axe, shovel and knife are common examples. The longer the wedge is, the greater the mechanical advantage. The mechanical advantage of a wedge can be found by dividing the length of either slope (S) by the thickness (T) of the big end. MA = Slope Length / Width

A Wedge

5. A screw is an inclined plane wrapped around a cylinder or a spiral inclined plane. The closer together the teeth are the greater the mechanical advantage. The mechanical advantage of a screw can be found by dividing the circumference of the screw by the pitch (lead) of the screw. A screw with a pitch of 1/8 and a circumference of 0.79 inches would produce a mechanical advantage of 6.3 (0.79 inches / 0.125 = 6.3).

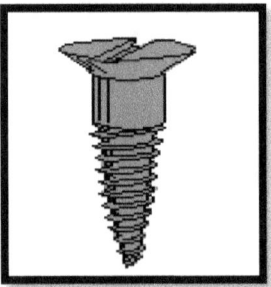
A Screw

6. A wheel and axle is made of two parts that turn together. When the wheel turns, the axle turns in the same direction. When the axle turns, the wheel turns in the same direction. The wheels of a car, the door knobs and the bottom of skateboards are all examples. The mechanical advantage of a wheel and axle is the ratio of the radius of the wheel to the radius of the axle.

A Wheel and Axle

Gears

A gear is a device in which toothed wheels engage to transmit motion between rotating shafts. If there are twice as many teeth on the output shaft, the input shaft will rotate twice as fast.

Gears

Comprehension Questions

1. List the six types of simple machines.
2. Study the simple machines and identify the one most effective for the jobs listed below:
   a. Getting water from a well
   b. Lift a heavy load on the back of a truck
   c. To get to a piano to the third floor of a building

3. Create a mnemonic device to help you remember the six simple machines.
4. Study the pictures and answer the questions which follow.

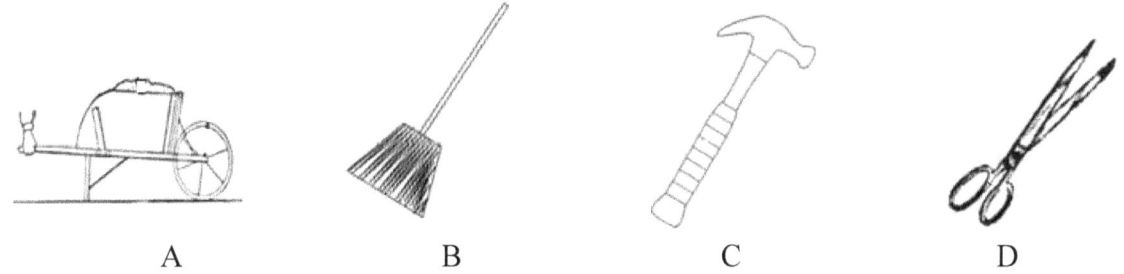

A   B   C   D

i. Identify the class of lever of each item above.
ii. What is the fulcrum of a lever?
iii. Name the other two parts of a lever.

5. Depending on what is in the **middle** of a lever determines what class of lever it is. What is in the middle of a:
a. Second class lever
b. Third class lever
6. Identify the six simple machines below.

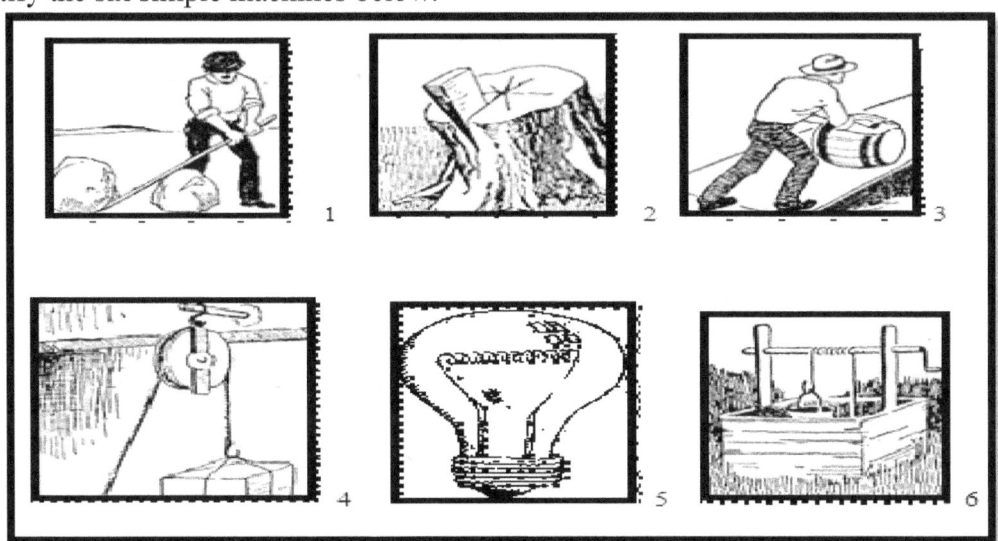

a.
   Name another simple machine in the same category as the machines shown above.

Mechanical Advantage of a Pulley System

The MA of a pulley is equal to the number of supporting ropes.

Interpret the diagrams to find the mechanical advantage of each pulley system.

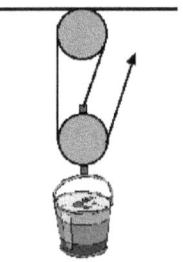　　　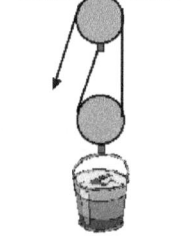　　　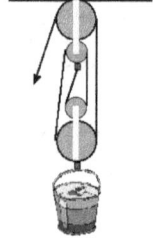

1. MA = _____　　　　2. MA = _____　　　　3. MA = _____

Mechanical Advantage of an Inclined Plane

To find the MA of an inclined plane, divide its length by its height.

MA = length / height

Interpret the diagram to answer the questions.

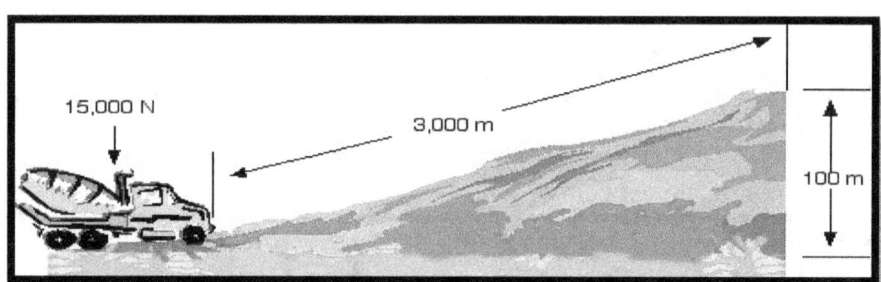

1. What is the height of the inclined plane? _____

2. What is the length of the inclined plane? _____

3. What is the mechanical advantage of the inclined plane? _____

4. How much effort force would be needed to push the dump truck up the mountain?

Forces

The S.I. unit for mass is the kg. **Mass** is the amount of matter in a substance. The pull of gravity on mass is a force called **weight**. (**Mass and weight are not the same**) That mass in kg x 10 gives us the weight or force due to the earth's pull on the substance. E.g. a 2 kg bag of flour has a force of   2 x 10 = 20N.

There are forces all around us. Any force applied can be categorized as a **push** or a **pull** or **twist**. If a force pushes, pulls or twists something, there are four main things that could happen:

- A motionless object might start to **move**
- The **pace** of a moving object might change
- The **direction** of a moving object might change
- The **shape** or size of an object might change

Machines are also used to produce large forces.

There exists a variety of forces: **friction, gravity, magnetism, air pressure, electrostatic and magnetic forces.**

As friction increases, motion decreases; as gravity increases, motion increases; as air pressure increases, movement decreases, as water pressure increases, movement decreases.

The S.I. unit related to forces is written as N or Newton. It was named after **Sir Isaac Newton**, a great scientist who formulated the laws of motion.

1. Friction

It is difficult to drag a heavy weight along a rough surface. This is because of the force called friction.

Friction is a force that slows down or stops the movement of objects. When one object rubs against another object, friction is the result. For example, when you rub your hands together they become warm. The kinetic energy from the rubbing changes to heat energy and your hands become warm.
Friction produces heat – striking a match
Friction causes things to make sounds – when you clap your hands
Friction causes wear and tear – that is why even metals wear down.

The rougher the surface, the more friction there is. For example: sand paper, concrete, match box. If there were no friction or a small amount, everything would keep sliding. Example: ice skating, glass and a bald tire. Without friction you would not be able to pick up anything because there would be no grip between your fingers and the object.

Friction can be reduced by **lubricating** the surface with **oil** or **grease**, or by making the surface **smoother**.   Using rollers or wheels between surfaces can also reduce friction. E.g. Putting a heavy load on wheels become much easier to move.

2. Magnetism

Magnetism is the invisible force of attraction or repulsion between substances like **iron, steel or nickel**. All magnets have a north pole and a south pole. A magnetic pole pushes or repels poles of the same kind, and pulls or attracts poles of the opposite kind. Like poles repel and unlike poles attract. Magnetism is strongest at the poles.

The region of force around a magnet or magnetic materials is called its magnetic field.

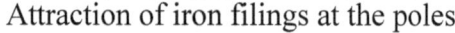

Attraction of iron filings at the poles

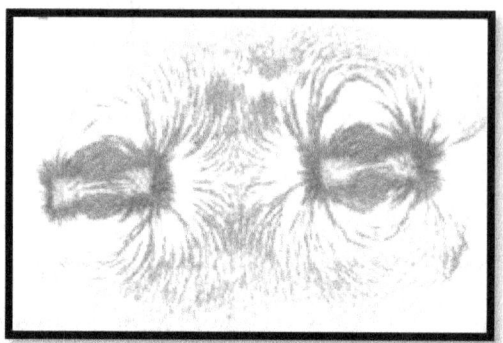

Magnetic field

When iron filings are placed around a bar magnet they are always arranged in the same kind of pattern. The lines are more crowded where the field strength is stronger and less crowded where the strength is weaker. These are called lines of force.

A magnet can be made by **stroking** or by **using electricity**. A wire coil that becomes a magnet when electric current moves through it is called an **electromagnet**. Electromagnets are found in telephones, earphones and machines that have electric motors.

The Earth acts like a giant magnet. If a bar magnet is supported so that it can turn freely, the magnetism of the Earth makes it turns until it points in a **north-south direction**.

A compass is an instrument that helps people use the Earth's magnetism to find direction. It has been discovered that one pole of a magnet that swings freely always points north.

3. Gravity

When you throw an object in the air it falls to the ground. The force that allows this to happen is the Earth's gravity. **Gravity** is the force that pulls everything down towards the center of the Earth and gives objects their weight.

Gravity pulls everything to the center of the Earth

The more gravity pulls on things, the greater the weight. The point at which the effect of gravity of an object seems to be concentrated is called the **center of gravity**. It is the point where the whole weight of the object seems to act. **There is NO gravity in outer space**.

4. Air pressure is the weight of tiny particles of air (air molecules) pressing down on the earth. Although air molecules are invisible, they still have weight and take up space. Since there is a lot of "empty" space between air molecules, air can be compressed to fit in a smaller volume.

When it is compressed, air is said to be "under high pressure". Air at sea level is what we normally feel to the point where we forget we are actually feeling it.

Air pressure is measured with an instrument called a **Barometer**. Air pressure is usually measured in *millibars* or in *inches of mercury*. Mercury in the Barometer adjusts until the weight of the mercury column balances the atmospheric force exerted on the reservoir. High atmospheric pressure forces the mercury higher in the column. Low pressure allows the mercury to drop to a lower level in the column. An aneroid barometer uses a small, flexible metal box called an aneroid cell. The box is tightly sealed after some of the air is removed, so that small changes in external air pressure cause the cell to expand or contract.

Air Pressure

A Barometer

5. Water pressure comes from the weight of the water overhead. The weight of the water depends on the height of the water column. It means that the deeper you are, the more the water above you weighs and the greater is the water pressure.

Near the surface of the water, the amount of water that is overhead is low, and therefore the water pressure is low. Deep underwater, there is a lot of water above you, and the water pressure is very high.

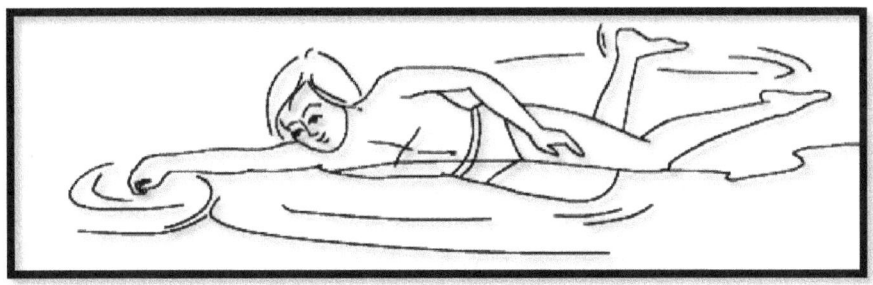

Water pressure

6. Electricity

Electricity is the movement of electrons through a conductor. The force needed to push the electrons through the conductor comes from a battery, generator or dynamo.

Comprehension Questions
1. What is mass?
2. What is the S.I unit for mass?
3. Define the term weight.
4. Calculate the weight or force of a block with a mass of 10kg.
5. How are forces categorized?
6. List four things that could happen when a force is applied to an object.
7. List one example of a force in animals.
8. Name FOUR types of forces.
9. What is friction?
10. List TWO things that could happen when friction is applied to an object.
11. Tell what happens to motion when:
    a. Friction increases
    b. Gravity increases
12. Explain how friction can be reduced.
13. Name the force which pulls objects to the center of the earth.
14. What is meant by the center of gravity?
15. You cannot pull your ring from your finger. Explain the steps you would attempt to remove the ring and give reasons for your answer.

Newton's Three Laws of Motion

First Law (**Law of Inertia**) – an object continues at rest or moves in uniform motion unless acted on by a force. (If an object is not moving, it will not start moving by itself. If an object is moving, it will not stop or change direction unless something pushes it.)

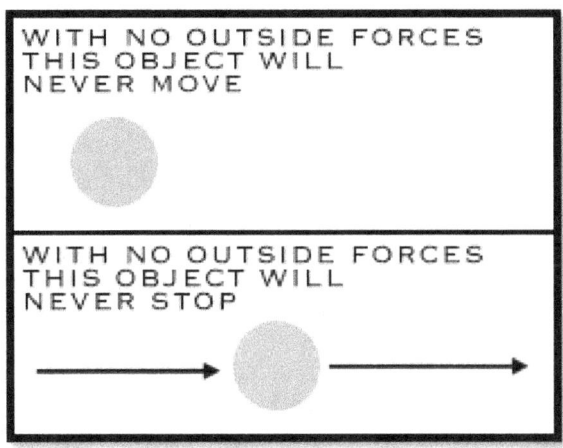

Example 1: if you drop a piece of popcorn on the floor, it would stay there until something moves it or picks it up.

Example 2: Resisting changes in your state of motion in an automobile while it is braking to a stop. The force of the road on the locked wheels provides the unbalanced force to change the car's state of motion, yet there is no unbalanced force to change your own state of motion. Thus, you continue in motion, sliding along the seat in forward motion. A person in motion stays in motion with the same speed and in the same direction unless acted upon by the unbalanced force of a seat belt. Seat belts are used to provide safety for passengers whose motion is governed by Newton's laws. The seat belt provides the unbalanced force that brings you from a state of motion to a state of rest. You could predict what would occur when no seat belt is used.

Second Law of Motion – an object continues in the direction of the force and at a speed proportional to the force. (Objects will move farther and faster when they are pushed harder.) The Second Law gives us an exact relationship between force, mass, and acceleration. It can be expressed as a mathematical equation:

$F = M\,A$ or Force = **Mass times Acceleration**

This is an example of how Newton's Second Law works:
Mike's car, which weighs 1,000 kg, is out of gas. Mike is trying to push the car to a gas station, and he makes the car go 0.05 m/s/s. Using Newton's Second Law, you can compute how much force Mike is applying to the car.

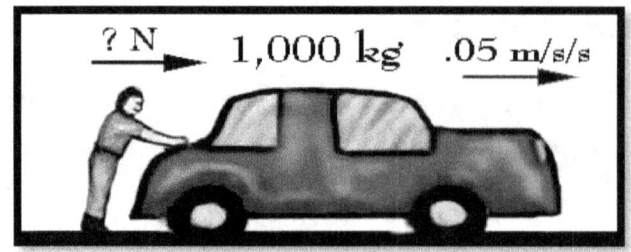

F=1,000 X 0.05

Answer = 50 Newton

Third Law of Motion – for every action there is an equal and opposite reaction. When an object is pushed in one direction, there is always a resistance of the same size in the opposite direction. Example: A boat on the water- as the boat pushes on the water, the water pushes on the boat with equal and opposite force which allows the boat to stay afloat. Without the opposite force of the water the boat would sink.

Equal and opposite reaction

When you hit a table with your hand, the table hits back with a force equal to the force you applied to it, resulting in pain or even damage to your hand. A rocket is able to lift off the launch pad because the acceleration imparted by the expanding exhaust is able to overcome the inertia of the rocket sitting on the pad. This is the same for a jet as it accelerates down the runway. The rocket continues to accelerate because the propellant used drives the mass of the rocket down, while the thrust of the engine continues unabated, leading to a real fast ride at burnout.

Useful Forces
a. friction between breaks and wheel helps a vehicle to stop its motion
b. sharpening a knife,
c. wind pressure for windmills to produce electricity,
d. gravity keeps things on earth,
e. water pressure used in reservoirs.

Force can be measured as the resistance offered. The greater the force applied, the greater the length of extension (coil or rubber).

Pressure
Pressure is the force exerted on a unit area of a surface. Pressure can be calculated using the following formula:
Pressure = Force / Area
It is more painful to be stepped on by a woman's stiletto heel than a broad square heel

Examples of Pressure Used in Everyday Life

A simple hydraulic press is used to raise a car in an auto shop. It consists of two large cylinders side by side. Each cylinder contains a piston, and the cylinders are connected at the bottom by a channel containing fluid. Valves control flow between the two cylinders. When one applies force by pressing down the piston in one cylinder (the input cylinder), this yields a uniform pressure that causes output in the second cylinder, pushing up a piston that raises the car.

In another variety of piston pump—the kind used to inflate a basketball or a bicycle tire—air is the fluid being pumped. Then there is a pump for water, which pumps drinking water from the ground. It may also be used to remove desirable water from an area where it is a hindrance, for instance, in the bottom of a boat.

Energy converted = work done
Work = Force x Distance
Force = Work / Distance
Distance = Work / Force

Comprehension Questions

1. Who was the scientist who gave us the Laws of Motion?
2. How many Laws of Motion are there?
3. What is another name for the first law of motion?
4. Which law explains why we need to wear seatbelts?
5. Which law says that force is equal to mass times acceleration (F=MA)?
6. Which law says that heavier objects require more force than lighter objects to move or accelerate them?
7. Which law explains how rockets are launched into space?
8. Which law says that for every action there is an equal and opposite reaction?

Sound Energy

Sound is a form of energy that helps us to hear. All sounds result from the vibration of particles in matter. Sound is produced when a force causes an object or substance to vibrate — the energy is transferred through the substance in a wave. The energy in sound is far less than other forms of energy.

Sound is made when something vibrates or moves back and forth. All sound is caused by a back and forth movement of materials. E.g. guitar strings, violins, drums, saxophones, whistles, vocal cords, etc.

A vibrating drum in a disco transfers energy to the room as sound. Kinetic energy from the moving air molecules transfers the sound energy to the dancers' eardrums. Kinetic (movement) energy in the sticks is transferred into sound energy.

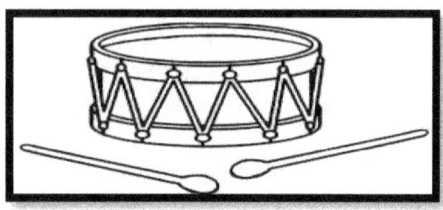
A drum can be made to vibrate

How does Sound Travel?

Sound cannot travel in a vacuum because there is no matter to vibrate. Sound needs matter to vibrate. Without this sound cannot be heard.

Sound is transmitted as sound waves through matter in any state, including gases (such as air), liquids (such as water), and solids (a door). Sound travels faster in solids than through liquids, because particles of a solid are more closely packed than liquids. Sound travels faster through liquids than through gases for the same reason. Therefore, sound travels fastest through solids and slowest through air because of the distance of particles from each other in each state.

Motion Waves

The ripples made when you throw a rock into a pond are similar to **motion waves**. When you talk or shout, your sound also travels in waves. These are **sound waves** but we cannot see them.

Motion Waves

When an object vibrates, it causes the molecules in the air to move. Therefore sound waves are created and travel outward from the source of the sound in all directions.

The Bell- jar Experiment

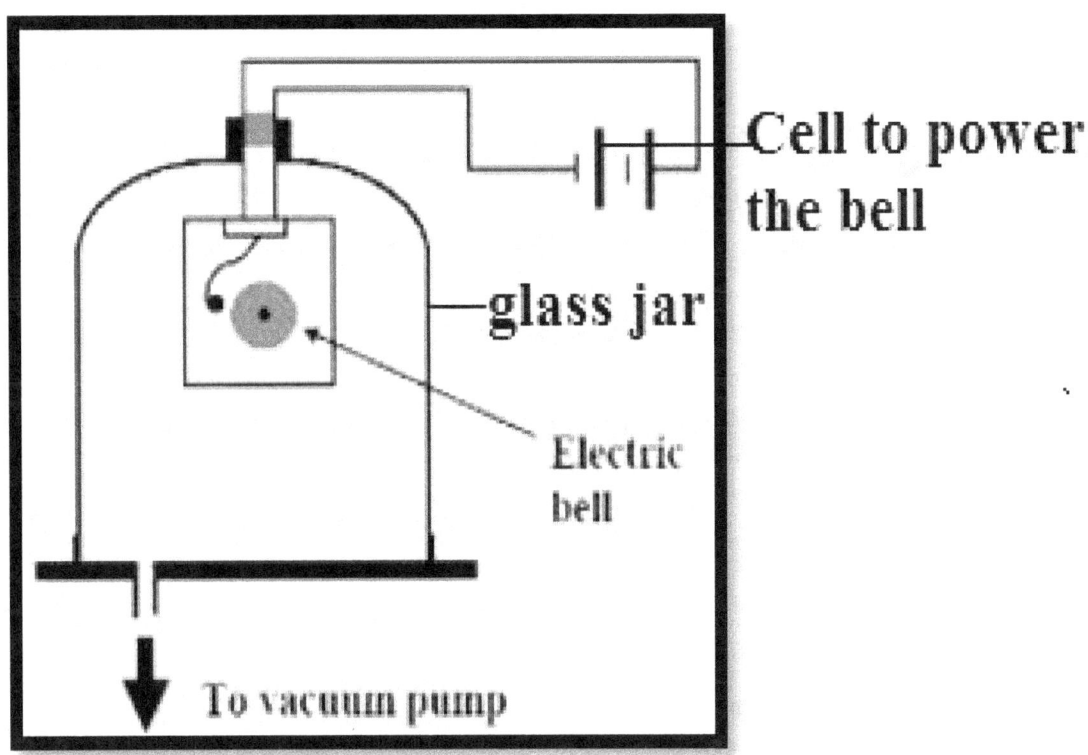

The diagram shows how the electric bell works

A student sets up the following experiment to investigate how sound travels through air. An electric bell was placed inside a bell-jar as shown in the diagram. The bell rang and it could be heard clearly. When the pump was switched on, it pumped the air out of the bell-jar and a vacuum was created. At that stage when the bell rang again it could no longer be heard but it could still be seen ringing. Vibrations could also be felt. This experiment demonstrates that sound is a form of energy.

Speed of Sound at Standard Temperature and Pressure (STP)

Speed of sound in wood   =   4,100 meters/second

Speed of sound in water   =   1,400 meters/second

Speed of sound in air   =   330 meters/second

Speed of sound in a vacuum   =   0 meters/second

Diagram of a Sound Wave

We can see a picture of a sound wave on the screen of an instrument called an **oscilloscope.** Transverse waves move side to side. Longitudinal waves move up and down.

Parts of a Transverse Wave

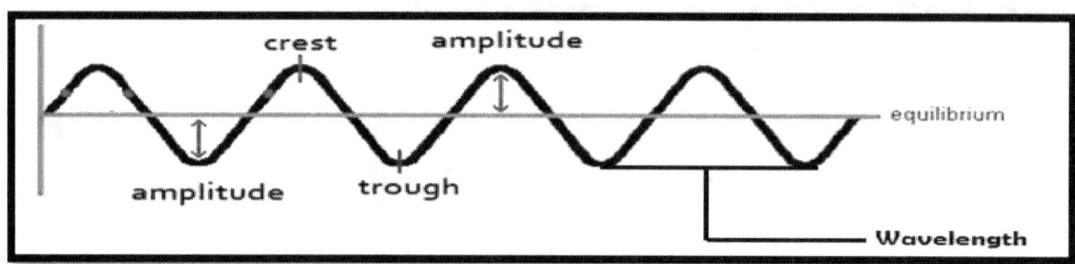

The diagram shows the parts of a wave

The line drawn through the center of the diagram represents the equilibrium or rest position of a string before the movement to create a wave.

Parts of a Wave

The **crest** of a wave is the point at the top of the wave's curve.
The bottom of the wave's curve is called the **trough.**
The distance from one crest to the next crest (or from one trough to the next trough) is called **wavelength.**

The **amplitude** is the measurement from the middle level of a wave to the crest or trough. The amplitude of the sound wave causes **loudness.** Loudness is measured in decibels.

The amount of energy carried by a wave is related to the amplitude of the wave. A high energy wave is characterized by a high amplitude; a low energy wave is characterized by a low amplitude.

Parts of a Longitudinal Wave

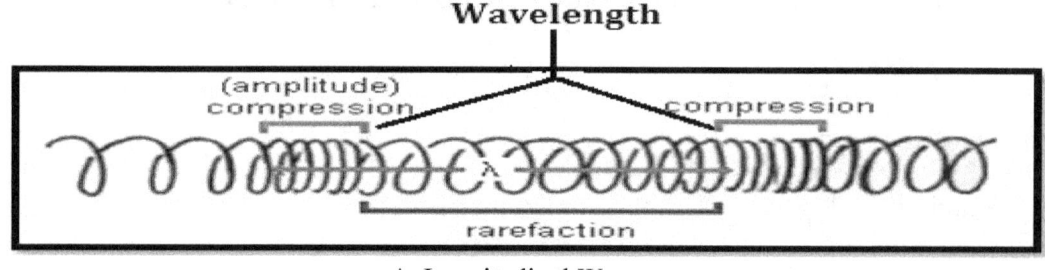

A Longitudinal Wave

Compression – area that is squeezed together; dark bands.

Rarefaction – area that is spread apart; light bands.

Amplitude – is measured from the equilibrium position to maximum compression; the length of one compression.

Wavelength – is measured from one compression to compression or rarefaction to rarefaction or any two successive identical points on the wave.

The Three Characteristics of Sound

Sound has three characteristics by which it may be recognized:

Tone (Quality)   **Pitch**   **Loudness**

Tone or Quality

Two sounds can have the same loudness and frequency but still sound different because of **quality or tone**. Some sounds are pleasant while others are a noise. Noise is merely unwanted or unpleasant sound. The quality of a sound depends upon the waveform.

A pleasant sound has a regular wave pattern. The pattern is repeated over and over. Noise on the other hand has an irregular wave pattern. The pattern is not repeated.

Pure Sound vs. Noise

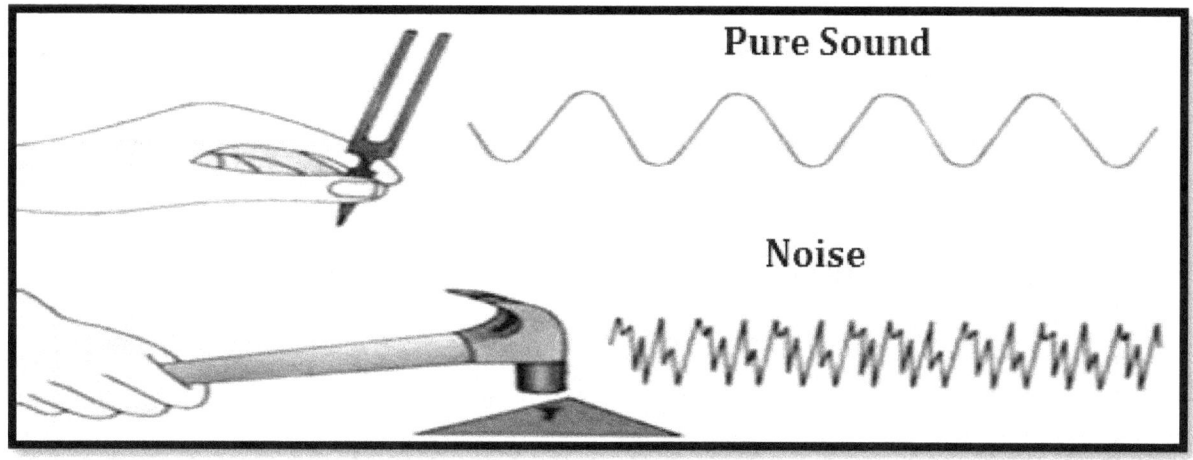

Pure sound vs. Noise

Pitch / Frequency

The number of waves that pass a point in one second is called **frequency**. The sensation of a **frequency** is commonly referred to as the **pitch** of a sound. Frequency is measured in hertz (Hz). 1 Hertz = 1 vibration per second. A measurement of one hertz means one full wave passes a point in one second. If twelve waves pass a point in one second, the frequency would measure 12 hertz. A high pitch sound corresponds to a high frequency sound wave and a low pitch sound corresponds to a low frequency sound wave. The shape of the sound wave changes if the sound becomes louder and has a higher pitch.

The diagram below shows two pressure-time plots, one corresponding to a high frequency (6 Hz) and the other to a low frequency (3 Hz).

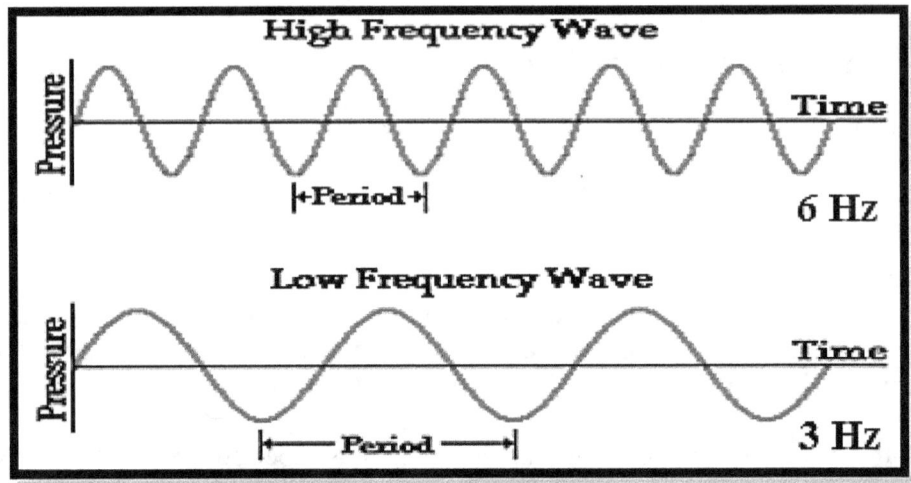

The diagram shows high and low frequency waves

Loudness / Amplitude

Loudness of a sound depends upon the **amplitude** of the sound wave. The more energy the sound wave has the louder the sound seems. You hear intensity as loudness.

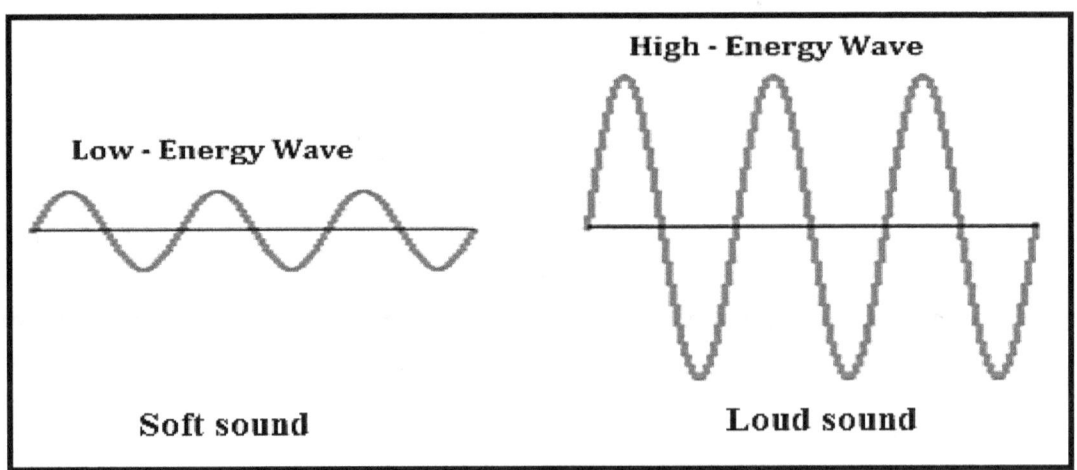

High and low energy waves

The loudness of a sound is a measure of the effect of the energy content of sound on the ear. It is related to the decibel (dB) which is a logarithmic scale used to measure the power of sound. The numbers in the diagram below represents the sound levels in decibels (dB).

Environmental Sounds Levels Measured in Decibels (dB)

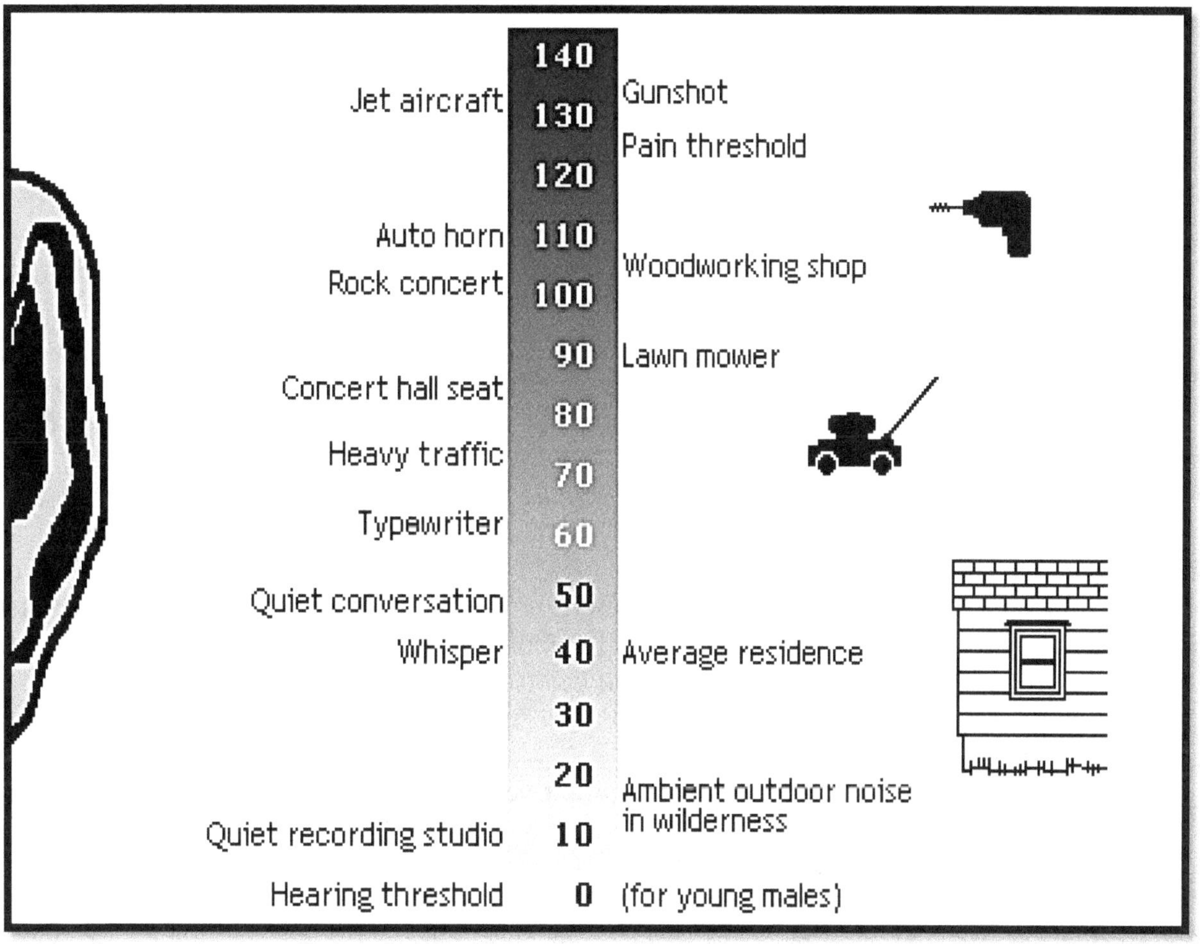

	140	Gunshot
Jet aircraft	**130**	
		Pain threshold
	120	
Auto horn	**110**	
		Woodworking shop
Rock concert	**100**	
	90	Lawn mower
Concert hall seat	**80**	
Heavy traffic	**70**	
Typewriter	**60**	
Quiet conversation	**50**	
Whisper	**40**	Average residence
	30	
	20	Ambient outdoor noise in wilderness
Quiet recording studio	**10**	
Hearing threshold	**0**	(for young males)

Various environmental sounds and decibels

Length of Column of Air and Pitch

Fill six bottles with water to different levels. Blow into each to see which gives a sound with the highest pitch.

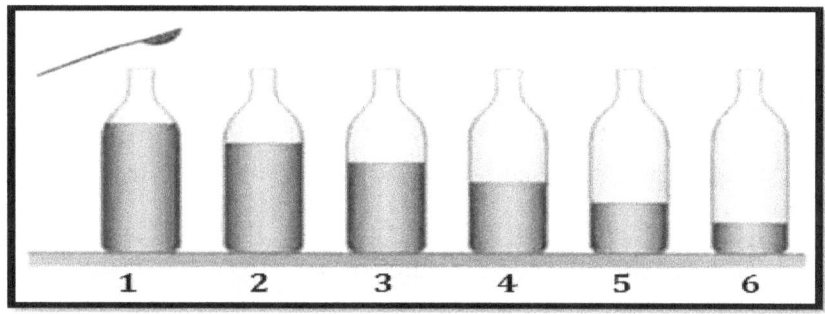

The longer the tube of air, the lower the pitch. This is why trumpets have a higher pitch than trombones.

Changing Pitch

Pitch varies with length and thickness (diameter).

E.g.

Pitch varies with length and thickness. The same length of strings with different thickness will produce different sounds. The thinner string has the higher pitch.

It is possible, however, to produce pitches with different frequencies from the same string. The four properties of the string that affect its frequency are length, diameter, tension, and density.

1. When the **length** of a string is changed, it will vibrate with a different frequency. Shorter strings have higher frequency and therefore higher pitch. When a musician presses her finger on a string, she shortens its length. The more fingers she adds to the string, the shorter she makes it, and the higher the pitch will be.
2. **Diameter** is the thickness of the string. Thick strings with large diameters vibrate slower and have lower frequencies than thin ones. A thin string with a 10 millimeter diameter will have a frequency twice as high as one with a larger, 20 millimeter diameter. This means that the thin string will sound one octave above the thicker one.
3. A string stretched between two points, such as on a stringed instrument, will have **tension**. Tension refers to how tightly the string is stretched. Tightening the string gives it a higher frequency while loosening it lowers the frequency. When string players tighten or loosen their strings, they are altering the pitches to make them in tune.
4. The **density** of a string will also affect its frequency. Dense molecules vibrate at slower speeds. The denser the string is, the slower it will vibrate, and the lower its frequency will be. Instruments often have strings made of different materials. The strings used for low pitches will be made of a more dense material than the strings used for high pitches.

Infrasound and Ultrasound

There is a lower threshold and upper threshold for sound. Below this threshold humans cannot hear and above this threshold will cause pain.

The human ear is capable of detecting sound waves with a wide range of frequencies, ranging between approximately 20 Hz to 20, 000 Hz.

Any sound with a frequency **below** the audible range of hearing (i.e., less than 20 Hz) is known as an **infrasound.**

Any sound with a frequency **above** the audible range of hearing (i.e., more than 20, 000 Hz) is known as an **ultrasound.**

Many other animals can detect a wide range of frequencies. Dogs can detect frequencies as low as 50 Hz and as high as 45, 000 Hz. Cats can detect frequencies as low as 45 Hz and as high as 85, 000 Hz. Bats are night creatures and must rely on sound echolocation movement and hunting. Bats can detect frequencies as high as 120, 000 Hz. Dolphins can detect frequencies as high as 200, 000 Hz. While dogs, cats, bats, and dolphins have an unusual ability to detect high frequencies, an elephant possesses the unusual ability to detect low frequency ranging approximately 5 Hz to 10, 000 Hz.

Uses of Sound in Modern Technology

The primary application of these very low frequency sound waves is in:

- Seismographs that detect and monitor earthquakes.
- Research on using infrasound as a weapon to disorientate the enemy.
- Various animals, such as the elephant, whale, rhinoceros and hippopotamus, use infrasound to communicate.

Ultrasound is used in sonograms to produce pictures of fetuses in the human womb, as well as to show other features in soft tissues without using x-rays.

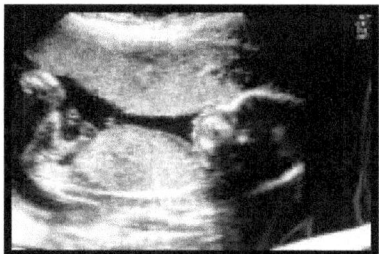 Fetal ultrasound

Ultrasonic cleaners are used to clean jewelry, lenses, medical instruments and industrial parts. These cleaners usually use frequencies around 30 kHz.
In industry, ultrasonic frequencies in the megahertz (MHz) range are often used to find flaws in materials, as well as to measure the thickness of objects where common measurements will not do.

Echoes

An echo is the reflection of sound when it hits a hard surface, such as a wall or cliff, the waves bounce back. Sound waves that bounce back is called an **echo**. Bats fly with their mouths open. They make noises and listen for the sound to bounce back - they listen for the echo. Bats are very good at hearing echoes, they depend on echolocation to hunt for food. When they hear the echo, they can tell where an object is, how big or little it is, and how far away it is. They can tell if the object is moving. They can tell what direction it is going. This is called **echolocation.** Echolocation is using sounds and their echoes to locate objects. If the object is a tree or a building, bats can keep from bumping into it. If the object is a bug, they can find it and catch it for food.

Echolocation in Bats

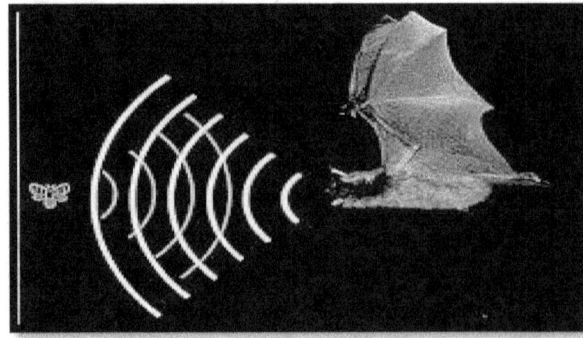

Bats use Echolocation

Echolocation in Dolphins

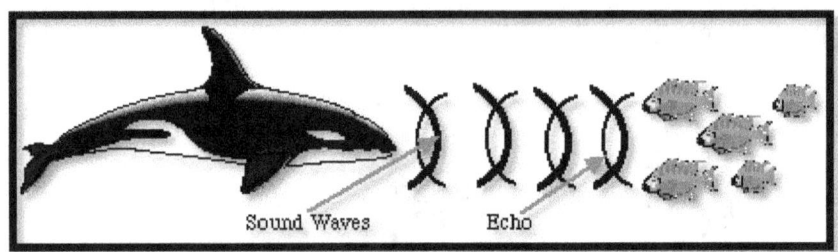

Dolphins, whales, seals, porpoises and some birds use echolocation

Effects of Loud Noises to the Human Ear – Noise Pollution

Noise pollution is defined as the production of unwelcome and displeasing sounds in the environment. Noise pollution affects humans and animals both in their behavioral and physiological health. Among the various sources of noise pollution are automobiles, aircraft and other transport systems. Other noise pollution causes, include industrial machinery, office equipment, vehicular horns, car alarms, sirens and audio speakers.

Cardio-vascular Issues

A noisy environment can be a source of heart problems and dramatic rise in blood pressure as noise levels constrict the arteries, disrupting the blood flow. A study was conducted where the heart rate of children staying in noisy surroundings was measured and it was found to be more than the heart rate of children living in less noisy environment. These sudden abnormal changes in the blood increase the likelihood of cardiovascular diseases in the long run.

Sleep Interruption

It is difficult to sleep comfortably when exposed to high decibel noise, which can affect your overall well-being. Noise can interrupt a good night's sleep, and when this occurs, the person feels annoyed, uncomfortable and extreme fatigue. This can considerably decrease a person's ability to work effectively.

Trouble Communicating

A noisy environment that produces more than 50-60 decibels does not allow 2 people to speak to each other properly. Hearing and understanding can become difficult and may lead to miscommunication.

Mental Health Problems

Having constant noise in the vicinity can lead to elevated stress levels as well as encourage violent behavior. The constant noise can also cause headaches, make people tense and anxious, and disturb emotional balance.

Hearing Problems

Loud noise tends to damage cells concerned with hearing (ear drum). Hearing impairment due to noise pollution can either be temporary or permanent. When the sound level crosses the 70 dB mark, it becomes noise for the ear. Noise levels above 80 decibels produce damaging effects to the ear. When ear is exposed to extreme loud noise (above 100 decibels) for a considerable period of time, it can cause irreparable damage and lead to permanent hearing loss.

Poor Cognitive Function

With regular exposure to loud noise, the ability to read, learn and understand decreases significantly over time. Problem solving capabilities and the ability to recall may also decline due to frequent bombardment of noise. Research has proved that children studying in a noisy environment tend to show relatively low cognitive function. The cognitive ability of children sent to schools that are in the close proximity of highways is less, in comparison to those learning in quieter surroundings.

Comprehension Questions

1. How are sounds made?
2. Explain how sound travels.
3. Consider the diagram below in order to answer A and B.

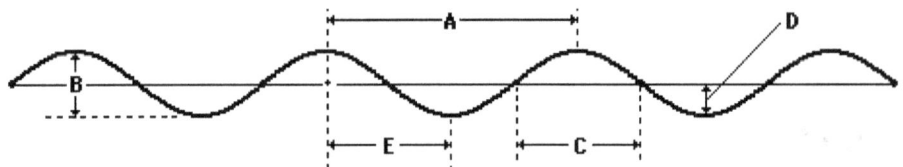

a. The wavelength of the wave in the diagram above is given by letter _____.
b. The amplitude of the wave in the diagram above is given by letter _____.

4. Indicate the interval which represents one full wavelength.

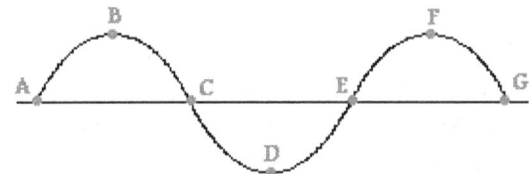

a. A to C

b. B to D

c. A to G

d. C to G

5. Distinguish the difference between frequency and speed.
6. Explain the correlation between pitch and frequency.
7. How is the loudness of sounds measured?
8. State the speed of sound through air at sea level.
9. Explain the following terms:

a. Echo
b. Infrasound
c. Ultrasound

Comprehension Questions

Study the diagram of the wave and answer the questions.

1. Identify the parts of the wave labeled A, B, C, D and E.

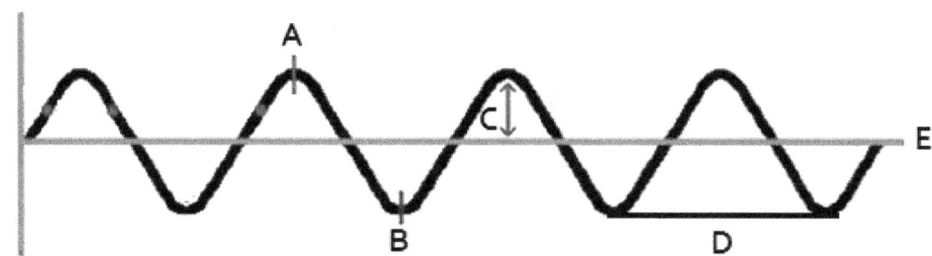

2. The diagrams below show the patterns made by four sound waves on an oscilloscope screen.

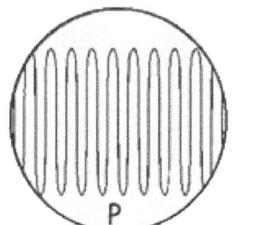 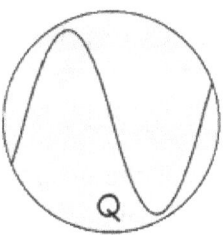 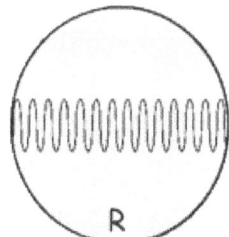 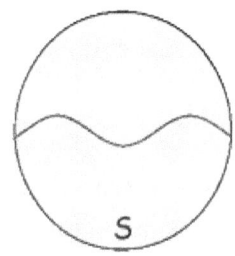

Write the letter of the sound wave that matches each of the descriptions below.

(a) a **loud** sound with a **low pitch**   _____

(b) a **quiet** sound with a **high pitch**   _____

(c) a **loud** sound with a **high pitch**   _____

3. State the THREE characteristics of sound.

4. Define the terms:

a. Infrasound
b. Ultrasound

5. Distinguish between pure sound and noise.

Activities
1. Manipulate a tuning fork to produce sound.
2. Blow across a candy wrapper to produce sound.

Endangered Species and the Effects of Urbanization

In order to accommodate the oversized human population, more and more lands are taken away from animals and plants due to **urbanization growth.** This leaves very little space for the natural habitats of many animals. With the limited land left, the food source will become limited. They have to compete for space and food in order to survive. Sometimes animals or plants do not adapt to the limited space so they die. This leads to a decrease in their population.

Threatened- Any animal or plant species whose numbers have been reduced to the point where they are **at risk** of becoming endangered.

Endangered species – Any animal or plant species whose numbers have been reduced to the point where they are **at risk** of becoming extinct.

Some examples of endangered animal species in The Bahamas are Hawksbill Turtles, Loggerhead Turtles, Iguanas, White Crown Pigeons, Bahamas Parrots, West Indian Flamingos, West Indian Whistling Ducks, Bahamas Hutias and the Bahamian Boa Constrictors.

Rare orchids are examples of endangered plant species.

Extinction – when all organisms of a species have died out, the species is termed extinct. Examples of organisms that have become extinct are dinosaurs and mammoths.

Biodiversity- the term biodiversity is used to describe a wide variety of species occupying a certain habitat. Every organism has its own intrinsic value.

Pollution is another huge factor causing animals and plants to become endangered. Animals need a clean habitat to survive. By eating our garbage unknowingly, they sometimes get poisoned or choked to death. Many become entangled and choked in six-packs bottle holders. Fish and birds get entangled in fishing lines and die. Toxic waste in the water system has caused a large number of fish to die out. By polluting the Earth, other animals and plants suffer.

Hunting and trading are other reasons that threaten the lives of many animals. People hunt and kill animals or plants for sport or for trading. Many of them do it illegally. When the rate of harvesting is greater than the survival, the population decreases.

The Role Overfishing / Overharvesting

Overharvesting or overfishing is defined as the excessive fishing of aquatic animals that include fish and shellfish. Overfishing can deplete some species of fish and marine invertebrates to very low numbers.   If the rate of harvesting is greater than the survival and maturation of juveniles to reproductive adults, the population decreases. Overharvesting or overfishing leads to an increase in the number of endangered species and can lead other species to the point of extinction.

If there is no control of harvesting and replenishing high demand foods, the population of these foods will decrease appreciably in the next ten years. Therefore these organisms that are used for food would become scarce. Examples are conch, grouper and pineapples.

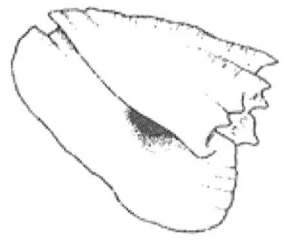
Conch

Nassau Grouper

Pineapples

Endangered Species of the Bahamas

Animal	Threats
Hawksbill Turtle (Reptile) 	**Commercial use**: • Hunted by man for its beautiful shell which is used to make tortoise shell combs, buttons, hair clips and jewelry. • Killed and stuffed as decoration. • **Habitat destruction:** Coastal development, and resulting pollution • Adult sea turtles are eaten by sharks. Young sea turtles are eaten by ants, crabs, dogs, raccoons, lizards, carnivorous fish and birds.
Loggerhead Turtle (Reptile) 	**Threats:** • **Commercial use**: Eggs are illegally collected by poachers for food. • **Habitat destruction:** Commercial development and resulting pollution. • **Natural threats**: Adult sea turtles are eaten by sharks. Young sea turtles are eaten by crabs, dogs, raccoons, carnivorous fish and.

Animal	Threats
Bahamian Rock Iguana (Reptile) 	**Threats:** • Wild hogs, feral cats and dogs. • Harvested by man for food. • Pet trade. • Natural disasters, such as hurricanes are a constant threat.
White Crown Pigeon (Bird) 	**Threats:** • Illegal hunting. • Loss of habitat. • Agriculture. • Deforestation.
Bahama Parrot (Bird) 	**Threats:** • Feral (wild) cats, feral boars, crabs and snakes. • **Heavy rains** during the nesting period can flood parrot nest holes, killing young chicks. • **Habitat loss**. • The **pet trade**.

Animal	Threat (s)
West Indian Flamingo (Bird) 	**Threats:** • The eggs can be trampled by wild donkeys and boars that roam freely where Flamingos live and nest. • **Illegal hunting:** It was hunted for its big, pink feathers used to decorate hats and other items. • The noisy disturbance of low flying planes of World War II drove the birds away.
West Indian Whistling Duck (Bird) 	**Threats:** • Excessive hunting. • Habitat destruction. • Predation. • Pollution. • Natural catastrophes: drought and hurricanes.
Bahamas Hutia (Mammal) 	**Threats:** • Consumed by local populations. • Predators included the extinct Alco, a Lucayan domesticated dog, and Chickcharnie owl. • Hurricanes. • Introduction of any predator.
Bahamian Boa Constrictor (Reptile) 	**Threats:** • Man - Many people kill these animals because of hysteria caused by ignorance and superstitious fear. • Habit destruction. • Collection for sale as pets.

Endangered Plant Species

Orchids have become endangered because of climate change, deforestation, changing land use and human activities. In addition, Orchids depend entirely on microscopic fungi in the early stages of their lives. Without the nutrients orchids get digesting these host fungi, their seeds often will not germinate and baby orchids will not grow. Not every fungus works for every orchid. If there is a mismatch, the fungus is useless. Purple orchids grow in the pine forest of the Bahamas. Some species of rare orchids are shown below.

Rare Orchids

Cattleya gaskelliana

Dendrobium X superbiens

Purple orchid

Miltoniopsis Herr Alexander

Lignum Vitae (Guaiacum Sanctum)

Lignum Vitae is the Bahamas national tree. It is also known as ironwood, because the lumber from the tree is so dense that it will sink in seawater. Its main feature is the blue / purple flowers at the tips which bloom twice a year. The Arawaks used the tree as a medicine to treat syphilis when it was introduced by the Europeans. This tree is considered threatened because it is very slow growing.

Lignum Vitae

Yellow Elder (Tecoma Stans)

The Yellow Elder is the national flower of the Bahamas. It is a yellow, trumpet-shaped flower that grows in clusters. The yellow elder attracts many bees, butterflies and hummingbirds.

Yellow Elder

Practices of Mankind that Cause Extinction

- Urbanization - Destroying the forest, the dwelling place for animals.
- Poor fishing practices e.g. overfishing, bleaching, long line fishing.
- Excessive hunting for food or other materials such as ivory from elephants.
- Polluting the air, land or water can cause disease or death.

Sometimes a change in temperature or a decline in number of species in the community can cause some species to be replaced by others. This gradual change in the composition of the Community is called a **succession**.

Effect of Urbanization Growth

As a result of urbanization growth, the habitat of small populations of species will be destroyed. This will lead to decrease in their population.

Protected Endangered Species

The population of the endangered and protected species will increase appreciably during the next ten (10) tears.

Conservation Methods

- Educate the citizens – the success of conservation efforts depends on the support of the people.
- Protection by law e.g. Crawfish closed season April 1st to July 31st inclusive. The closed season for grouper (December to February) is announced annually. People should refrain from eating protected species except during the legal season.
- Prevent pollution of the air, water and land.
- Establishment of national parks and marine reserves to protect endangered Species.

National Parks

A national park is an area that is selected to protect:

- Pristine areas that are beautiful or of scientific interest

- Habitat or breeding grounds of endangered species

- Places of historic worth

- Special examples of vegetative or geological zones

Importance of National Parks

- They provide safe places for scientific research and education

- They protect resources for future generations

- They protect entire habitats and the biodiversity that relies on the habitats

- They are one of the greatest legacies to be left for future generations

National Parks of the Bahamas

To date, there are 27 national parks found throughout the Bahama Islands and many more are expected in the near future. Some National Parks are listed below.

Abaco National Park

It was established 1994 and is the major habitat of the Bahama Parrot.

Black Sound Cay National Reserve

It is located off Green Turtle Cay in Abaco. It is made up of mangrove vegetation which is an important habitat for waterfowl and other birds.

Blue Holes National Park

It is located in Andros. The caves house many unusual and unique cave fish and invertebrates, some of them are not found anywhere else in the world.

Bonefish Pond

It is located in New Providence and is a coastal wetland where teachers and students can learn about the different marine organisms.

Conception Island National Park

It is located on San Salvador and is a sanctuary for migratory birds, sea birds and green turtles.

Crab Replenishment Reserve

It is located in central Andros and is a habitat and sanctuary for land crabs.

Exuma Cays Land and Sea Park

Is the first marine fishery reserve established in the Caribbean.

The Need for Biodiversity

Biodiversity is the term used to describe a wide variety of species occupying a certain habitat. Every organism has its own intrinsic value.

Conservation Organizations

- Bahamas Marine Mammal Survey (BMMS)
- Bahamas National Trust (BNT)
- Bahamas Reef Environment Educational Foundation (BREEF)
- Dolphin Encounters
- Friends of the Environment
- The Bahamas Environment Science and Technology Commission (BEST)
- The Nature Conservancy (TNC)

Comprehension Questions

Read each clue and for each one identify the endangered species and one conservation method in place to protect it.

1. Illegal hunting, loss of habitat, agriculture and deforestation have become a problem for my feeding grounds, especially inland hardwood forests. I have a brilliant white forehead and crown to the head.

2. My bill is black and legs are dark with a greenish tint. My numbers decline due to excessive hunting, habitat destruction, predation, pollution and natural catastrophes such as drought and hurricanes.

3. My nests are in the ground which makes me vulnerable to predation by feral (wild) cats, feral boars, crabs and snakes. Heavy rains during the nesting period can flood my nest holes, killing my young chicks. I have two toes facing forwards and two facing backwards.

4. I am a prize for my beautiful shell which is used to make shell combs, buttons, hair clips and jewelry. Coastal development and resulting pollution also contributes to my decline.

5. I am not to be confused with the others of my kind. I build my nests directly on the ground which makes me vulnerable to a number of predators. My eggs can be trampled by wild donkeys and boar. In the past I was also hunted for my feathers that were used to decorate hats.

6. My numbers decline because of excessive hunting, habitat destruction, predation, pollution and catastrophes such as drought and hurricanes.

7. Many people fear me, but I am harmless. I do not have venom so my bite is not poisonous. I am persecuted because of ignorance and superstitious fear. My decline is also due to habitat destruction and collection for sale as pets.

8. I am consumed by local inhabitants; I am also vulnerable to natural disasters such as hurricanes.

9. I was named by fishermen. My eggs are illegally collected by poachers for food. I have few enemies because of my tough shell; however, crabs, dogs, raccoons, some fish and birds eat my young after hatching.

Activities

Effect of Over Fishing on Endangered Species Activities

1. Add a fixed number of items to a container (e.g. 5 every 10 minutes). Use fishing gear to extract as many items while fishing for 10 seconds every minute.
2. Find out about TWO species that have become extinct in the past ten years.
3. Find out about TWO species that have become endangered in North, Central or South America in the past five years.
4. Write an infomercial about them.
5. Host a forum or prepare a comic strip drama or prepare a pamphlet or brochure flyer soliciting the support of peers, relatives or the community to practice conservation measures to protect small population species.
6. Select an endangered plant species of the Bahamas and complete the chart with information about the plant.

Air

All living things need air to live. Plants and animals that live in water also need air to live. Cold water contains air in the form of bubbles. Air bubbles will collect on plants also. Sea animals breathe out air, which appears as bubbles in the water. Air is in water, soil, plants and animals. Things that seem empty have air inside.

Air is made up of particles of matter. Molecules of air give air its mass. The weight of air is due to the force of gravity acting on its mass.

Properties of Air
1. Air is real
2. Air occupies space
3. Air has mass and weight
4. Air exerts pressure

Physical Properties of Air
Air is colourless,
Air is odourless,
Air has mass,
Air has weight,
Air has density

Air Occupies Space – Has Volume

The experiment below proves that air occupies space.

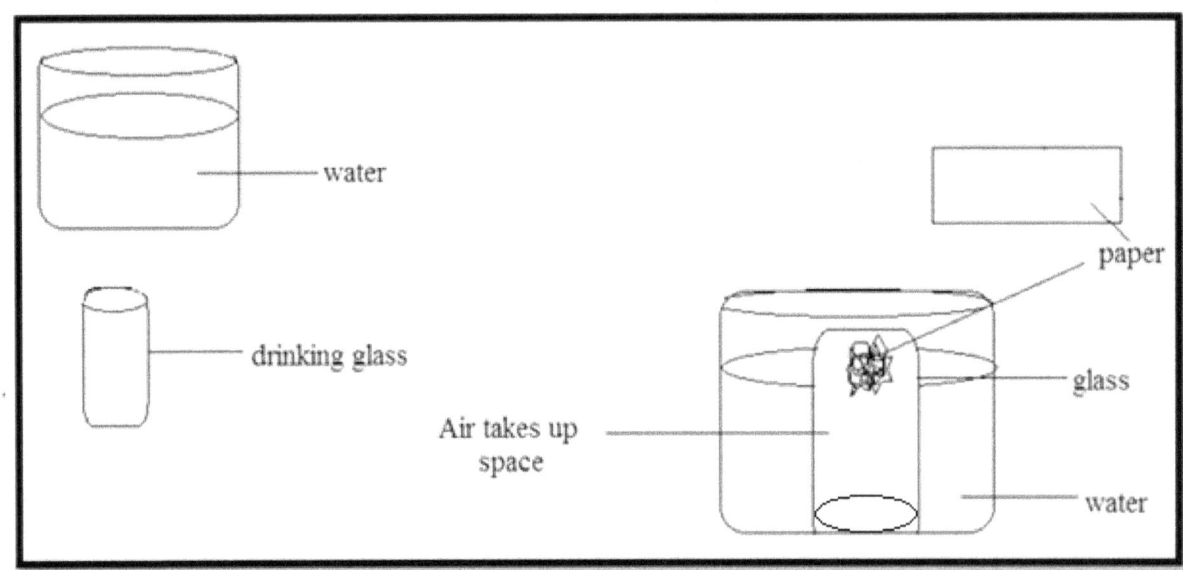

Experiment to prove that air takes up space

Two types of matter cannot occupy the same space simultaneously (at the same time). When the glass is submerged in the water, the compressed air serves as a barrier between the paper and the water and the paper remains dry.

Air has Mass

The two balloons are of equal weight with air. However, when the white balloon is deflated that side of the stick will rise as it is lighter. The balloon with air is heavier. This proves that air has mass / weight.

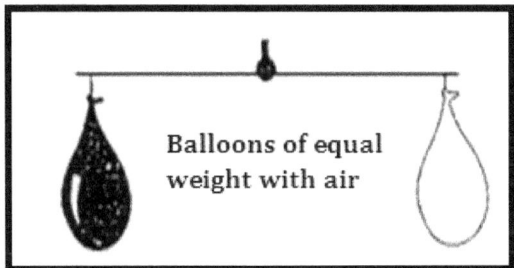 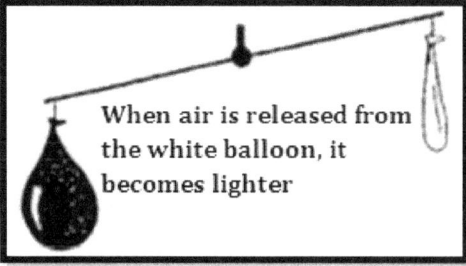

Experiment to prove that air has weight

Air Exerts Pressure

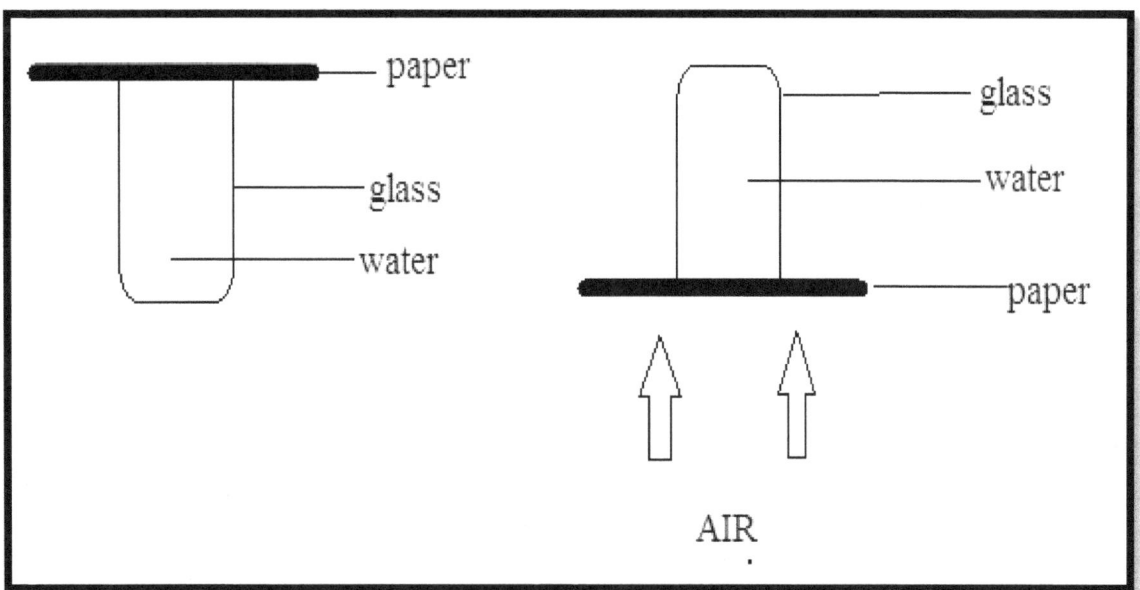

The hard paper will not fall off because air exerts pressure / force in all directions.

Composition of Air

Air is a mixture of gases, mainly nitrogen and oxygen. A group of gases known as Rare or Noble gases occupy a small part of the air along with water vapour which varies, and carbon dioxide. Air also contains suspended dust, spores and bacteria.

Composition of Air by Percentage

Nitrogen 78%
Oxygen 21%
Carbon dioxide 0.03 -0.4%
Noble gases 0.9 -1.0%
Water vapour varying quantities

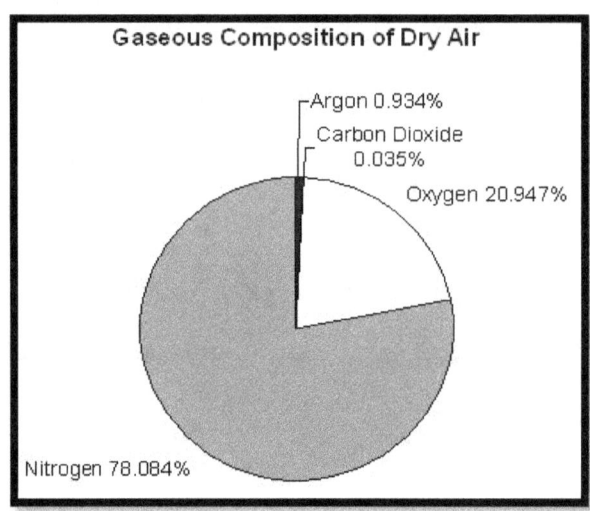

Pie Chart Showing Approximate Percentage of Oxygen in Air

Chemical Composition of Air

Name	Chemical Symbol	% by Volume	Importance
Nitrogen	N_2	**78.084 %**	used as fertilizers in plants
Oxygen	O_2	**20.9476%**	used in burning things and respiration
Argon (RG)	Ar	0.934%	used in light bulbs
Carbon Dioxide	CO_2	0.0314%	used for photosynthesis in plants
Neon (RG)	Ne	0.001818%	used in bright colourful lights
Methane	CH_4	0.0002%	used as fuel to make heat and light (Liquefied natural **gas)**
Helium (RG)	He	0.000524%	one of the lightest gases, is used for filling balloons
Krypton (RG)	Kr	0.000114%	used in light bulbs
Hydrogen	H_2	0.00005%	used in gas tanks for cutting, welding and as a lifting **gas** for airships. It is highly flammable and can be dangerous to use.
Xenon (RG)	Xe	0.0000087%	used in light bulbs
Water vapour	H_2O	Varies	used to make rain clouds in the sky

RG = Rare gas

Density of Air

Density is the mass of matter divided by the volume. To find the **density of air**, the mass of a sample of air is measured and compared to the volume it occupies. Density of air decreases as altitude increases. The force of gravity is greatest near Earth's surface therefore more air molecules are found there.

As altitude increases, gravity decreases therefore less air molecules are present.

Comprehension Questions

1. Explain why air is important.

2. Blow into a balloon and tie the top.
a. Feel the balloon with your hands. What is inside the balloon?
   b. Untie the balloon, squeeze and hold it near your cheek. What did you feel?
   c. Can you squeeze out all of what was inside?
   d. Did you see what came out?
   e. What colour was it?

3. List SIX properties of air.
4. Which gas occupies the following proportions of the atmosphere?

   i.   20 %
   ii.   78 %
   iii.   0.03 %
   iv.   0.93 %

5. Write the symbol for each of the following:

   i.   Water vapour
   ii.   Helium
   iii.   Krypton
   iv. Xenon

6. State the importance of the following:

   i.   Hydrogen
   ii.   Methane
   iii.   Neon

7. a. Define the term density.
b. Explain what happens to the density of air as altitude increases.
8. What happens to gravity as altitude increases?
9. Describe an experiment to prove that:
   a. Air takes up space
   b. Air exerts pressure
10. There is NO air in outer space. List FOUR organisms that you would not find in outer space.

Mathematics Skill
Pie charts and bar graphs

Use a table showing the percentage composition by volume of the various gases in air. Draw a pie chart and a bar graph.

Oxygen Gas - O_2
Oxygen is one of the most important gases in the air. Without oxygen, most plants and animals would die. About 20 % of the air is made up of oxygen (O_2). When plants make food in the process of photosynthesis, oxygen is made as a waste product and released in the air. Oxygen is taken from the air when animals breathe in (inhale). This oxygen is used in the process of respiration.

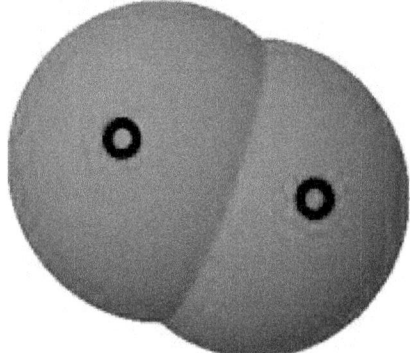

An Oxygen Molecule

Physical Properties of Oxygen
1. Colourless gas
2. Odourless gas
3. Slightly soluble in water.
4. Density of oxygen is slightly greater than density of air.

Chemical Properties of Oxygen
1. Oxygen is a very reactive element and reacts with most elements forming oxides.
2. Oxygen supports combustion.
3. Oxygen will allow many substances to burn in it forming oxides.

An oxide is a compound of an element with oxygen. **Examples of Oxides:** magnesium oxide, copper (cuprous/cupric) oxide, sulphur dioxide, sulphur trioxide, calcium oxide, iron (ferrous/ferric) oxide.

Uses of Oxygen
1. Oxygen is used as a respiratory aid (combustion of food to produce energy).
   - in high altitude flying and climbing
   - in deep sea diving
   - in hospitals

Uses of Oxygen

Oxygen at High Altitude   Oxygen in Deep Sea Diving   Oxygen in Hospital

2. Oxygen facilitates the burning of rocket fuels.

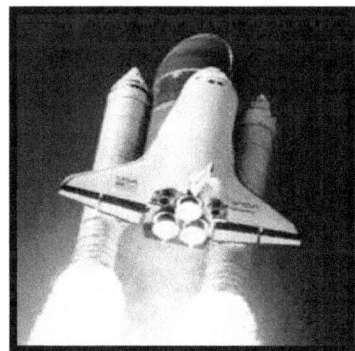

Use of Oxygen in Rocket Fuel

3. Oxygen is used in welding and cutting metals.

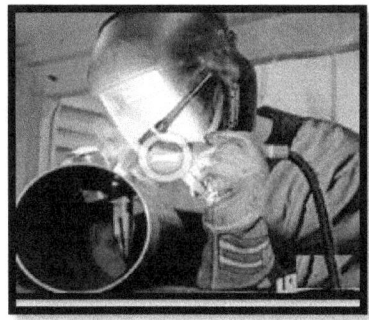

Use of Oxygen in Welding

Combustible and Non-Combustible Substances

A combustible substance is one that burns in air or oxygen.
Examples are wood, leaves, paper, wax, hydrogen gas, ethanol, methane, propane, butane, benzene

A non- combustible substance is a material that will not ignite, burn, support combustion, or release flammable vapors when subjected to fire or heat. Examples are metals, concrete, fiberglass, ceramic tiles, asbestos, etc. Non-combustible materials are useful in the construction industry.

Laboratory Preparation of Oxygen

1. Oxygen gas can be prepared by heating a compound called potassium permanganate. This has to be done carefully because potassium permanganate is explosive. Potassium permanganate which is a purple crystalline solid, decomposes without fusing on heating, forming a black powder consisting of a mixture of potassium manganate and manganese dioxide while releasing oxygen. As the gas is created and collected, it will displace the water from the container. The volume of gas can be determined by the amount of water that was displaced by the oxygen gas.

The Chemical reaction is:

$2 KMnO_4 \quad\quad\quad\quad ==> \quad\quad K_2MnO_4 \quad\quad + \quad\quad MnO_2 \quad + \quad O$

Potassium permanganate ==> Potassium manganate + Manganese dioxide + oxygen

Preparation of Oxygen Gas Potassium Permanganate

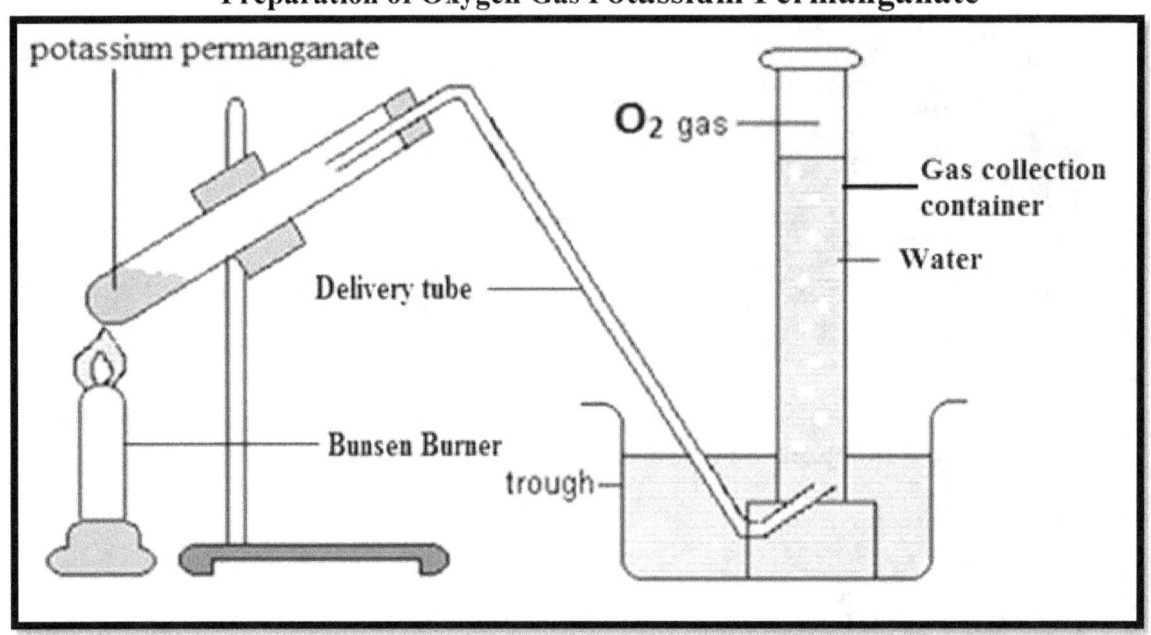

Preparation of Oxygen using Potassium permanganate

2. Oxygen can also be prepared by the catalytic breakdown or decomposition of hydrogen peroxide. The apparatus is set up as shown. Hydrogen peroxide is added slowly to the flask of Manganese Dioxide. Manganese dioxide is used as a catalyst to speed up this reaction and results in the production of oxygen **gas**. The gas is collected over water and collected by the upward displacement of air. As the gas is created and collected, it will displace water from the container. The volume of gas can be determined by the amount of water that was displaced by the gas.

Preparation of Oxygen Using Manganese Dioxide

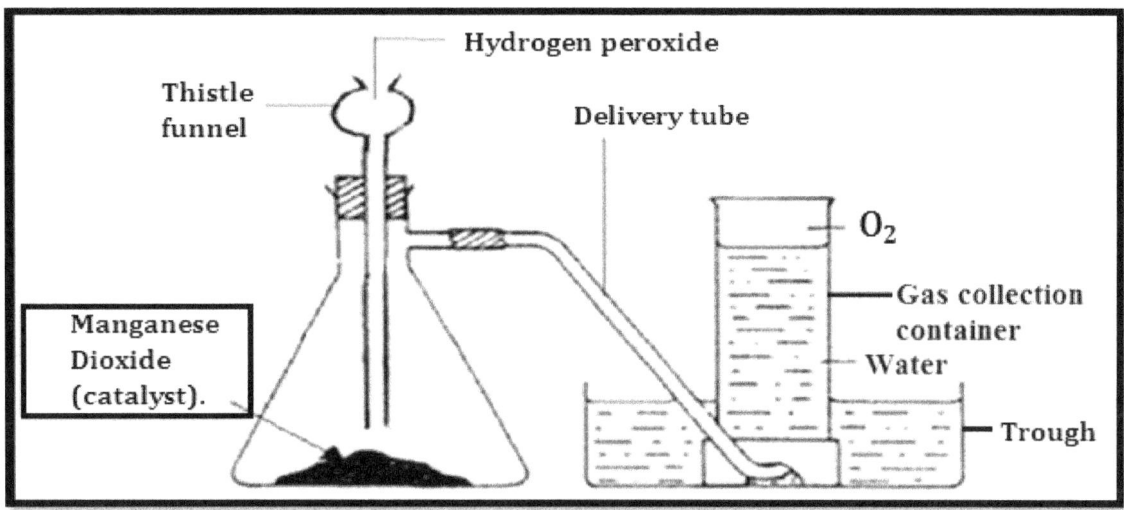

Oxygen prepared in the laboratory using manganese dioxide and hydrogen peroxide

The chemical reaction is: Catalyst (manganese dioxide)

$$\longrightarrow \quad \text{Hydrogen peroxide} \quad = \quad \text{Water} \quad + \quad \text{Oxygen}$$

$$2H_2O_2 \quad = \quad 2H_2O \quad + \quad O_2$$

The Need for Air-tight Conditions

Air-tight conditions are precautions taken during an experiment to prevent air from passing through a container prematurely. If O_2 is released too quickly, the rubber stopper will explode out. Air contains oxygen in a significant proportion which would affect combustion. Vaseline should be applied around the place where the delivery tube fits into hole in rubber stopper and where the rubber stopper meets the neck of the flask. This prevents gas from entering or escaping.

Test for Oxygen
Oxygen relights a glowing splint. It supports combustion.

Positive Test for Oxygen
To test for Oxygen gas, light a splint (piece of wood) and blow it out so it is just glowing. Place the splint into a test tube of oxygen. The splint will burst into flame, the glowing splint burns brightly indicating the gas is oxygen.

Positive Test for Oxygen Gas

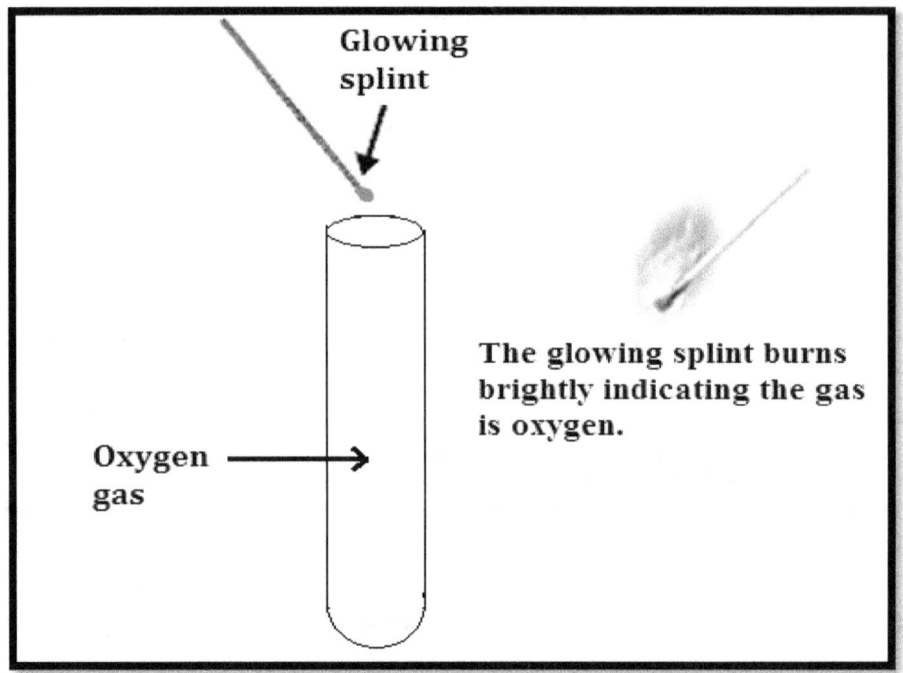

Positive test for oxygen

Comprehension Questions

1. Where does the oxygen in the air come from?
2. The formula for oxygen gas is
3. Oxygen makes up about _____ of the air.
4. What are THREE chemical properties of oxygen?
5. What are three physical properties of oxygen?
6 a. Define the term oxide.

   b. List TWO examples of oxides.
7. How is oxygen taken from the air?
8. Explain THREE uses for oxygen.
9 a. What is a combustible substance?

   b. List THREE examples:

   c. List THREE examples of non-combustible substances:
10 a. Explain why it is necessary to have air tight conditions.

   b. What procedure would you follow to achieve air tight conditions?

11. Draw a diagram to show how oxygen can be made in a laboratory using Potassium Permanganate. Draw in pencil and label in blue or black ink at the end of the line. Please ensure that the line touches the structure being labeled. Use a ruler to draw all lines.

12. The formula for a molecule of this substance is K₂MnO₄.
a. Find the number of atoms in the molecule shown above.
b. Identify and name any two elements in this molecule

13. Oxygen can also be produced in a laboratory using Manganese dioxide and hydrogen peroxide. What is the purpose of Manganese dioxide?

14. Explain how you can test to see if a gas is oxygen.
15. Write the chemical formula for:

a. Hydrogen peroxide

b. Water

Activity
Design a gadget that might safely produce oxygen using:
Reactants chamber, passage to collecting container, air-tight conditions. (Write lab report)

Carbon Dioxide Gas - CO_2

Carbon dioxide is made up of the elements carbon and oxygen. The formula for carbon dioxide is **CO_2**. Carbon Dioxide makes up about 0.03% or 0.04% of the air.

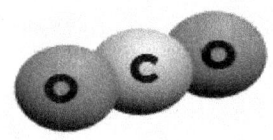

$$C + O_2 \dashrightarrow CO_2$$
Carbon Dioxide Molecule **CO_2**

Physical Properties of Carbon Dioxide
1. Colourless gas
2. Odourless gas
3. Denser than air
4. Slightly soluble in water

Chemical Properties of Carbon Dioxide
1. It does not support combustion.
2. It turns limewater milky.
3. It turns damp blue litmus paper pink as carbonic.

Carbon Dioxide is always being put back into the air when:
    i.   Animals breathe out.
    ii.   Dead plants and animals decay or rot.
    iii.   Fuels are burned.

Uses of Carbon Dioxide

1. It is dissolved in water under pressure to make **"fizzy" drinks**.

Carbon dioxide makes the fizz in drink

2. It is used in **fire extinguishers** to put out fires. It does not support combustion.

Carbon dioxide in fire extinguisher

3. It is used in baking to **help bread rise**. Yeast produces carbon dioxide which makes bubbles and causes the bread to rise. The yeast is killed in baking.

Carbon dioxide produced by yeast helps bread to rise

4. It is taken from the air and used in the process of photosynthesis to help plants make food.

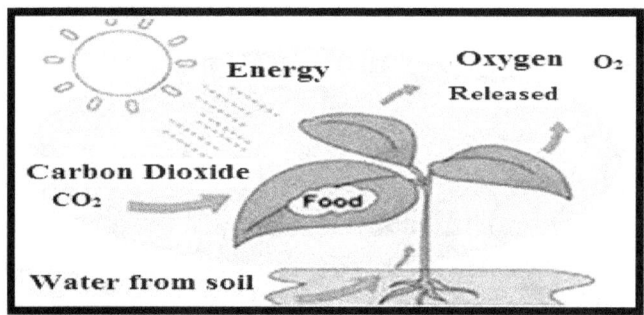

Carbon dioxide in the process of photosynthesis

Dry Ice - Carbon dioxide is easily changed to a solid called dry ice. It sublimes and is useful for cooling things. Dry ice evaporates rather than melt, so there is no messy water if it is used in a cooler after a picnic. Caution must be taken because **Dry ice** is much colder **than regular ice**, and can burn the skin similar to frostbite.

Laboratory Preparation of Carbon Dioxide

Carbon dioxide is prepared in the laboratory from the reaction between dilute hydrochloric acid and marble chips (a form of calcium carbonate).

In the laboratory, carbon dioxide is usually prepared by the action of dilute hydrochloric acid on marble chips (a form of calcium carbonate). Marble chips are taken in a round-bottomed flask and dilute hydrochloric acid is added through the thistle funnel. A glass jar filled with water is placed on the beehive shelf over the mouth of the delivery tube. The gas produced by the reaction comes out of the delivery tube and displaces water in the gas jar. This method is called collecting gas over water.

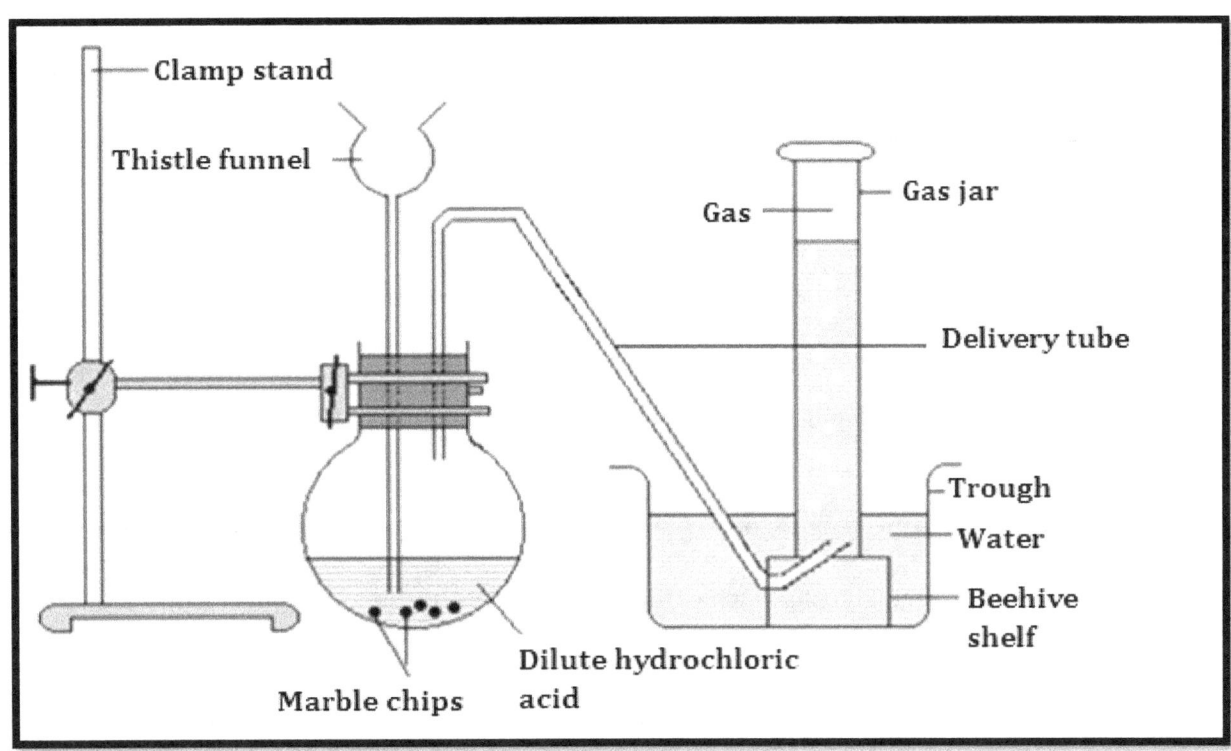

Carbon dioxide preparation in the laboratory

The chemical equation of the reaction is:

$$CaCO_3(s) + 2HCl(aq) \longrightarrow CaCl_2(aq) + H_2O(l) + CO_2(g)$$

| Calcium carbonate | Hydrochloric acid | | Calcium chloride | Water | Carbon dioxide |

Test for Carbon Dioxide

1. Carbon dioxide turns limewater milky (formation of calcium carbonate).

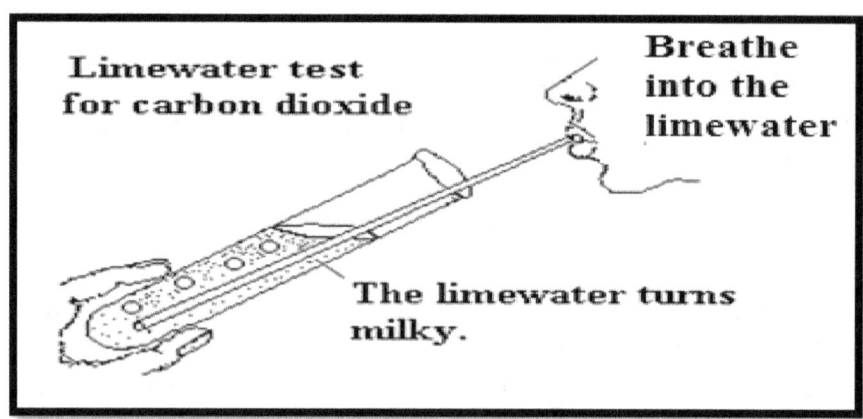

The diagram shows the test for carbon dioxide by breathing into limewater

When carbon dioxide collects into limewater, the limewater turns milky

The diagram shows what happens when carbon dioxide is collected over limewater

2. Carbon Dioxide Extinguishes a Glowing Splint.

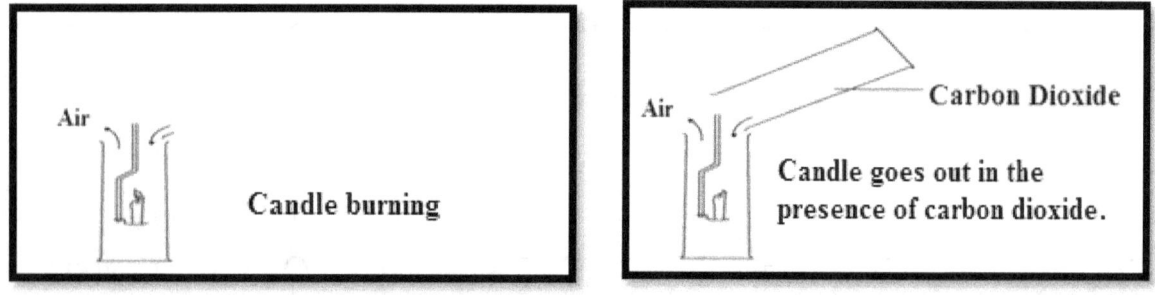

Carbon dioxide puts out a lighted splint. Carbon dioxide does NOT support combustion.

Remember:

Dilute sulphuric acid should not be used in this preparation of carbon dioxide as it forms a white insoluble layer of calcium sulphate around the marble chips. This coating does not allow the marble chips to come in contact with the acid. As result, the reaction stops.

Carbon dioxide is prepared in the laboratory from the reaction between dilute hydrochloric acid and marble chips (a form of calcium carbonate).

Comprehension Questions

1. Carbon dioxide is made up of _____ and _____.
2. The formula for carbon dioxide is
3. Carbon dioxide makes up about _____ of the air.
4. What are three properties of carbon dioxide?
5. Carbon dioxide is put back into the air when:
6. Carbon dioxide is taken from the air when:
7. List and explain four uses for carbon dioxide.
8. Explain how you can test to see if a gas is carbon dioxide.
9. Briefly explain how carbon dioxide can be made in a laboratory.
10. Write the chemical formula for:
a. Hydrochloric acid
b. Calcium carbonate (Marble chips)

11. Explain why:
a. dry ice is better to use than regular ice.
b. carbon dioxide is a good gas to use in fire extinguishers.

12. Draw a diagram to show how carbon dioxide can be made in a laboratory.

13. Explain in your own words the term "Global Warming."

14. Design a fire Extinguisher - A container with a substance which easily puts out fire. Container must be easily accessed and content easily discharged.

Activities

Laboratory preparation of carbon dioxide
Write a lab report
Draw conclusions on the presence of carbon dioxide based on the test for carbon dioxide.
Lime water become milky in the presence of carbon dioxide.

Carbon

Organic substances are substances made by living organisms. They are large molecules and always contain carbon. Glucose is an organic substance. Organic matter, when roasted turns black indicating the presence of carbon. Soot is a form of carbon that is formed.

Differences between Blue and Yellow Flames

Blue flame	Yellow flame
Very hot	Not very hot
Forms no soot	Forms soot
Can be noisy	Quiet
Has three zones	Has four zones
Steady flame	Unsteady flame
Carbon is completely burnt.	Un-burnt carbon glows yellow
More heat	Less heat
Less Light	More light

Carbon is the one element on which all life depends. Carbon can be found in many forms. It exists in the atmosphere as carbon dioxide (CO_2) and on the Earth as fossil fuels. Coal, natural gas and petroleum are all compounds which are based on carbon. Carbon exists in forms called allotropes which are pure forms of carbon. The most common is diamond. The other pure carbon allotrope is graphite.

Graphite

Graphite is a common form of carbon found in countries such as Sri Lanka, U.S.A., and Germany. It is a shiny black crystalline substance which is reasonably soft to the touch. Analysis has revealed that graphite consists of layers of carbon atoms formed in six rings. These layers can slide over each other, resulting in a soft, slippery feeling. Graphite is used as a solid lubricant, in pencils, and as a moderator in nuclear reactors. In its crude form, graphite is located in veins within metamorphic rocks. Geographically, large deposits of graphite are found in China.

Diamond

Diamonds are found in igneous rocks and come mainly from South Africa. A diamond's structure is that of a "lattice", its atoms are arranged in a perfectly cubical shape, similar to common table salt. It is made up of carbon. Each carbon atom in a diamond is surrounded by four other carbon atoms and connected to them by strong covalent bonds. Diamond is the hardest substance known to man and is used for other things besides jewelry. Diamonds occur in a variety of colors including, red, pink, yellow, purple, blue and green.

Diamonds are used for:
1. Cutting tools
2. Drilling
3. Grinding
4. Polishing procedures

Diamonds:

have a high index of refraction,

are high in luster - sparkle

disperse light high

are excellent insulators.

All of these properties help to make diamond the world's most popular gemstone.

Both diamond and graphite are made entirely of carbon. The way the carbon atoms are arranged in space, however, is different. The differing properties of carbon and diamond arise from their distinct structures.

Properties of Diamond and Graphite

Colour, hardness, source (where they are found) and time to form are all related to the physical properties and uses of graphite and diamond.

Diamond	Graphite
Hardest mineral known to man.	A soft metal and greasy to touch.
An excellent electrical insulator.	A good conductor of electricity.
Very abrasive.	A very good lubricant.
Usually transparent, occur in a variety of colors including, red, pink, yellow, purple, blue and green.	Opaque (black in colour).
Rare and expensive, world's most popular gemstone.	Cheap and abundant.
Non-malleable.	Malleable.
Used in jewelry.	Used in pencils (lead); steelmaking.
Crystallizes in the Isometric system	Crystallizes in the hexagonal system.

Diamonds and graphite are two crystalline allotropes of carbon. Diamond and graphite both are covalent crystals. But, they differ considerably in their physical properties.

Diamonds

Graphite – lead and coal

Importance of Bacteria in the Carbon Cycle

When organisms die the carbon compounds in their remains are broken down by decomposers like bacteria and fungi. Carbon dioxide is released into the atmosphere; hence, carbon dioxide is recycled in the ecosystem.

Carbon in the Carbon Cycle

Trees provide oxygen and remove carbon dioxide from the atmosphere. Removal of too many trees causes an increase in carbon dioxide in the atmosphere.

Factor That Would Interrupt the Carbon Cycle

Deforestation

Deforestation is the intentional removal or destruction of trees from the forest. Removing the forests interferes with the Earth's ability to remove carbon dioxide from the atmosphere, so the amount of CO_2 remains high. Carbon dioxide is a greenhouse gas. A greenhouse gas stops some of the heat from escaping into space which would increase the temperature of the atmosphere.

Comprehension Questions

1. What is an organic substance?

2. What happens to an organic substance when it is roasted?

3. What is soot?

4. Explain THREE differences between a blue flame and a yellow flame.

5. Explain the ways in which carbon exists on earth.

6. Tell THREE interesting points about graphite.

7. Tell FIVE interesting points about diamond.

8. Explain ONE similarity and ONE difference between graphite and diamond.

9. Explain the importance of bacteria in the carbon cycle.

10. Define the term deforestation and explain the negative impact on the environment.

The Carbon Cycle

The carbon cycle is the pathway along which carbon travels in nature. Carbon dioxide in the air is the source of most of the carbon in living things. Carbon Dioxide is taken from the air by green plants in the process of photosynthesis. It is replaced into the air by all living organisms when they exhale and by burning.

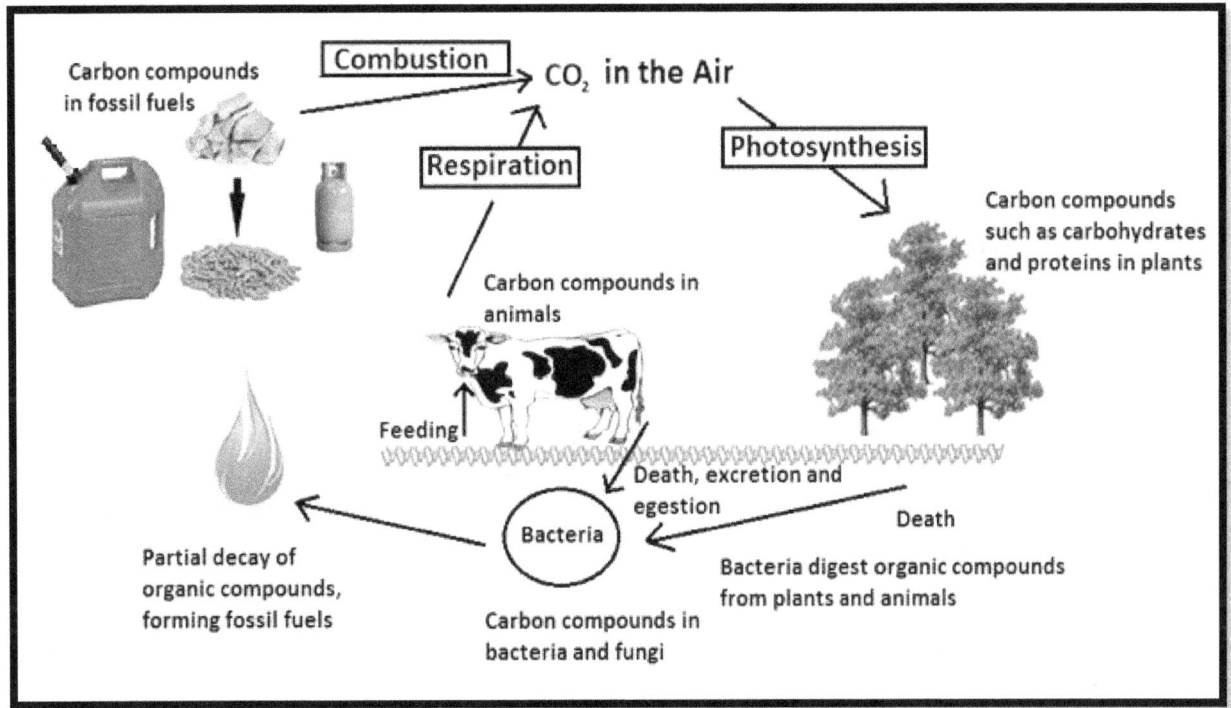

The diagram shows the Carbon Cycle

Photosynthesis Uses Carbon Dioxide

Only about 0.04 % of the air is carbon dioxide. Plants use carbon dioxide from the air in their food making process called photosynthesis. The carbon from the carbon dioxide becomes part of the food molecules. Plants only photosynthesize during the daylight, because photosynthesis needs energy from sunlight.

Respiration Produces Carbon Dioxide

All living things need energy. They get their energy from food. Energy is released from food in a process called respiration. During this process, carbon dioxide is produced as a waste product. This is where the carbon dioxide comes from when you breathe out.

All living things respire. So, all living things produce carbon dioxide. Even plants produce carbon dioxide. But during the day, plants take in carbon dioxide for photosynthesis faster than they make it in respiration. In the night, plants give out carbon dioxide faster during respiration. They produce oxygen as a result of photosynthesis. In the daytime plants take in carbon dioxide and at night they give it out. Trees are therefore useful for both reasons.

Burning (Combustion) Produces Carbon Dioxide

When things burn they react with oxygen in the air. The fuels which we burn all contain carbon. The carbon reacts with oxygen in the air to produce carbon dioxide.

Fossil fuels, such as coal, oil and gas, were formed from the remains of plants and bacteria. The plants took carbon dioxide from the air. So the fuels contain carbon. The oceans, and other bodies of water, absorb some carbon from the atmosphere. The carbon is dissolved into the water.

Climate Change Due to Excess Carbon Dioxide in the Atmosphere

The buildup of carbon dioxide in the atmosphere in recent decades is causing **global warming**. The carbon dioxide behaves like the glass of a greenhouse. It lets the Sun's rays through to the Earth's surface but stops heat from escaping. This causes the trapping of heat on the Earth which results in global warming.

Increasing temperatures caused by climate change will make the water of the oceans expand and rise; ice melting in the Antarctic and Greenland will also cause the sea level to rise. Sea levels could rise by as much as 25 to 50 cm by 2100. Greater sea levels will threaten the low-lying coastal areas. Many areas of land will be in danger from flooding; causing people to leave their homes. Low lying areas in cities will be greatly affected by the rising sea.

Changes in weather will affect many crops grown around world. Crops such as wheat and rice grow well in high temperatures, while plants such as maize and sugarcane prefer cooler climates. Changes in rainfall patterns will also affect how well plants and crops grow. The effect of a change in the weather on plant growth may lead to some countries not having enough food. Many people could be affected by hunger.

Changes in the climate will change the weather patterns and will bring more rain in some countries, but others will have less rain. The dry areas will become drier and wet areas could become wetter.

Safe Practices to Reduce Carbon Dioxide Level in the Atmosphere

- Stop burning trash
- Plant trees
- Reduce use of electricity
- Use cars less frequently
- Regular servicing of motor vehicles. Vehicles release carbon dioxide into the atmosphere from burning fuel.

Carbon as the Main Source of Energy

- Coal is cheaper but produces soot (carbon) and carbon dioxide.
- Propane produces carbon dioxide, more expensive – less soot.
- Electricity is cleaner but more expensive, carbon dioxide at the plant / production level
- Formulate a hypothesis on the effect of a mandatory decrease in the number of vehicles would have on the health of people on a densely populated island

Oxygen – Carbon Dioxide Balance

1. If the rate of removal of oxygen by respiration and its rate of replacement by photosynthesis is about the same.
2. If the rate of removal of carbon dioxide by photosynthesis and its rate of replacement by respiration is about the same.

Are Trees More Useful as Oxygen Producers or as Carbon-Dioxide Absorbers?

During the day trees take in carbon dioxide for photosynthesis faster than they make it in respiration. In the night they give out carbon dioxide faster during respiration. They produce oxygen as a result of photosynthesis. Trees are useful for both reasons.

Respiration and Photosynthesis Compared

The process of respiration is the reverse of photosynthesis and vice versa. The exception is, in respiration energy is also produced.

Photosynthesis

Carbon Dioxide + Water ⟶ Glucose + Oxygen
(Raw Materials)   (Products)

Respiration

Glucose + Oxygen ⟶ Carbon Dioxide + Water + Energy
(Raw Materials)   (Products)

The products of photosynthesis are the raw materials for respiration and the products of respiration are the raw materials for photosynthesis.

Comprehension Questions

1. What is the carbon cycle?
2. What percentage of the air is carbon dioxide?
3. How is carbon dioxide removed from the atmosphere?
4. Explain the term photosynthesis.
5. Explain TWO ways in which carbon dioxide may be replaced in the atmosphere.
6. Explain the following terms:
a. Combustion
b. Respiration
7. How are fossil fuels formed?
8. List TWO examples of fossil fuels.
9. Explain what happens to the remains of dead organisms.
10. Explain TWO differences between respiration and photosynthesis.

11. Explain what happens when there is too much carbon dioxide in the atmosphere.
12. Explain ONE negative practice which results in increased levels of carbon dioxide in the air.
13. List THREE safe practices to reduce carbon dioxide level in the atmosphere.
14. Do you think trees are more useful as oxygen producers or as carbon-dioxide absorbers?

Activities:

1. Use a crucible or tongs to roast seeds.
2. Design a filter to remove un-burnt carbon (soot).
3. Use a yellow flame to heat water in a beaker and observe the soot on the beaker).
4. Use Bunsen burner to show differences between blue and yellow flames.

Other Activities:

1. Debate- Present points in a debate on the effects of carbon dioxide emissions on the greenhouse effect.

2. Design and conduct an investigation to determine the level of awareness of global warming in the community. Conduct a survey using questionnaires and interviews.
3. Design and conduct an investigation to show the extent to which people are utilizing one safe practice to reduce carbon dioxide level in the atmosphere. Conduct a survey using questionnaires and interviews.
4. Advocate peers, relatives or the community to practice energy conservation measures. Target group(s) through brochures, meetings, soap box etc.

5. Host a meeting, write a song/rap for a target group.
6. Formulate a hypothesis on creating oxygen – carbon dioxide balance in nature.
7. Write a paragraph explaining whether trees are more useful as oxygen producers or as carbon dioxide absorbers.
8. Activity: Follow instructions to prepare recycled paper.

Layers of the Atmosphere

The **atmosphere** is the layer of gases that surrounds the earth and makes conditions suitable for living. The air around you is part of the atmosphere. As you recall from the components of the atmosphere, the most common gas is nitrogen. Nitrogen makes up about 78 % of the atmosphere. The most familiar component however, is oxygen which makes up about 21 % of the atmosphere.

The atmosphere is divided into layers based on temperatures. These differences are a result of the way the sun's energy is absorbed as it travels through the atmosphere. Some layers contain gases that easily absorb the sun's energy, while other layers do not. The layers that absorb the sun's energy are warmer than the other layers. This is why the various layers have different temperatures. The five layers of the atmosphere are the troposphere, stratosphere, mesosphere, thermosphere and exosphere.

Layers of the Atmosphere

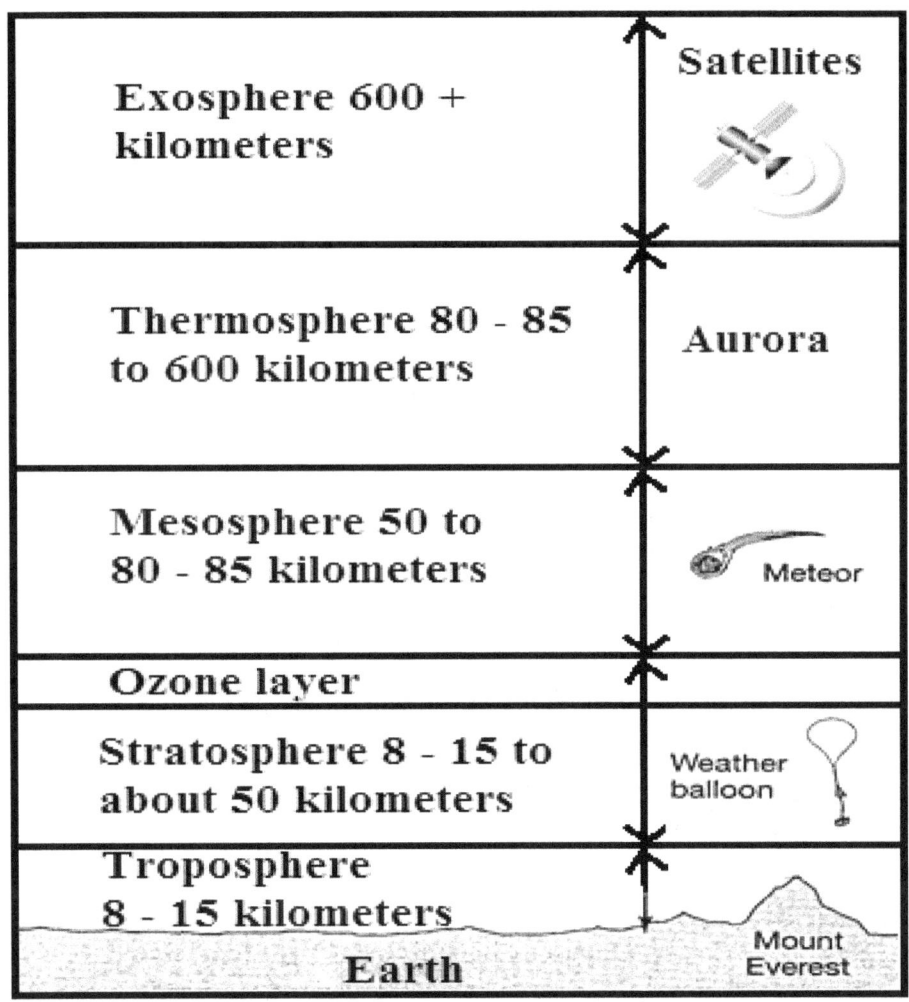

The diagram shows the layers of the atmosphere

Exosphere- is the outermost layer of the atmosphere. This is where satellites and space shuttles orbit. Beyond this layer is outer space where the air pressure is extremely low because there are so few molecules floating around.

Thermosphere- is beyond the mesosphere. In this layer, the temperature increases with altitude. An important part of the thermosphere is the **ionosphere**. This layer in the thermosphere is where gas particles are electrically charged due to being hit by the solar energy coming from the sun. The ionosphere plays an important role in making radio communications possible. **Auroras** are also found in the ionosphere. Auroras or the **northern lights** are brilliant displays of colors in the night sky.

Mesosphere – is the middle layer of the atmosphere. In this layer, radio waves are reflected to Earth. Just like the troposphere, temperatures drop as altitude increases, making this the coldest layer of the atmosphere. The mesosphere protects the Earth from meteoroids. They burn up as they fall towards earth in this layer of the atmosphere.

Stratosphere – is above the troposphere. The air is very thin and contains very little moisture. It is extremely cold in the lower part of the stratosphere. At the top of the stratosphere, there is a layer of **ozone**. Ozone is a form of oxygen essential to our survival. This **ozone layer** helps to protect the earth from the sun's ultraviolet (UV) radiation. The ozone absorbs the UV radiation, keeping it from getting down to the surface of the Earth where it can harm living organisms. In this layer, the temperature increases as altitude increases due to the ozone molecules absorbing ultraviolet radiation from the sun. Only the highest clouds (cirrus, cirrostratus, and cirrocumulus) are in the lower stratosphere.

Troposphere – the layer closest to the ground. It contains It contains almost all of Earth's carbon dioxide, water vapour, air pollution, weather and clouds. People, plants, animals and insects also live in the troposphere. The temperature is warmer at the surface of the earth and decreases as you go higher in the troposphere.

Comprehension questions

1. What is the atmosphere?
2. Why do different layers of the atmosphere have different temperatures?
3. What two gases are most common in our atmosphere?
4. Which layer of the atmosphere do you live in?
5. What might you find in the exosphere?
6. Which layer of the atmosphere makes radio communication possible?
7. This is the coldest layer of the atmosphere. _____
8. Where would you find ozone?
9. What is ozone and why is it important?
10. As you move up through the mesosphere, the temperature

_____

Soil

Soil covers much of the land on Earth. It is made up of minerals (rock, sand, clay, silt), air, water, and organic material (matter from dead plants and animals). Soil provides a substrate for plants to anchor their roots, a source of food for plants, and a home for many animals e.g. insects, spiders, centipedes, worms, burrowing animals, bacteria, and many others.

Humus is a dark-brown organic component of soil that is made up from decomposed plant and animal remains and animal excrement. Humus improves the water-retaining properties of soil, adds nutrients and makes it more workable.

Soil scientists describe soil types by how much sand, silt and clay are present. This is called texture. It is possible to change the texture by adding different substances in the correct proportions. Changing texture can help in providing the right conditions needed for plant growth.

Soil Particles

1. Sand is made up of large soil particles
2. Clay is made up of small particles
3. Silt is made up of medium particles

Types of Soil

Clay Soil: Has small air spaces. It therefore has a high water retention quality. Clay soil can hold a lot of nutrients, but does not let air and water through it well. It is necessary to drain clay soil frequently, for improving its texture. The soil becomes unmanageable during rainy season, as it becomes 'sticky'. On the other hand, during drought, it becomes 'rock solid'. (See the comparison of soil particles in the diagram to the right.)

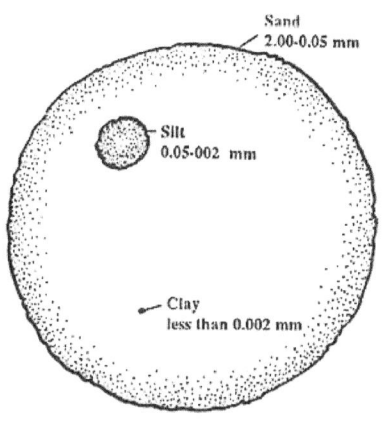

Sandy Soil: Have the largest particles in the soil. When you rub it, it feels rough. This is because it has sharp edges. Sand does not hold many nutrients. It is light and dry in nature. It does not have moisture content and warms up quickly in the spring. Plants rooted in sandy soil need to be watered frequently.

Loam Soil: Loamy soil is called the perfect soil or garden soil. It has a combination of all the various types in the correct proportions. It is suitable for any and every kind of crops. Loam soil has the best characteristic of all. It has high nutrient content, warms up quickly in summers and rarely dries out in the dry weather. It has become the ideal soil for growing crops.

Comprehension Questions

1. What makes up the contents of soil?
2. Define 'Humus'.
3. How would you change the texture of soil and give the three particles that make up soil.
4. Name the four (4) types of soil.
5. State the differences between the types of soil.
6. Sketch a diagram to represent sand, silt and clay soils.

Porosity of Soil

Porosity refers to the pores or spaces in the soil. The porosity of soil depends on the size of the soil particles. Soil with larger particles will have larger air spaces and allow water to pass through more quickly than soil with small particles. As the porosity increases, the drainage and leaching rate also increases.

Large Soil Particles   Small Soil Particles

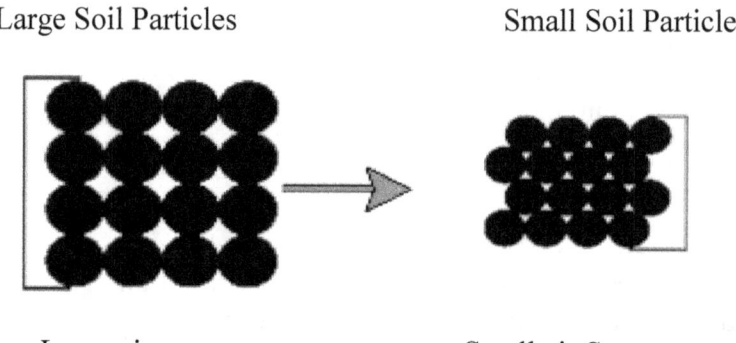

Large air spaces   Small air Spaces

When water moves into soil pores (spaces), it replaces air which is pushed out. This air rises as bubbles.

Volume of air = volume of soil + volume of water – final volume

Percentage = volume air / volume soil x 100

Soil Permeability

Soil permeability is the ability of water and air to travel through the open spaces, or pores, between grains of soil. Permeability generally refers to the top 5 feet of soil, but some soils may have multiple permeability.

The size and shape of soil pores determine soil permeability. Coarsely textured soils, including sand and gravel, generally have larger soil particles and have high soil permeability rate. More water will drain from soil with larger spaces than soils with small spaces.

Diagrams to Demonstrate Soil Permeability

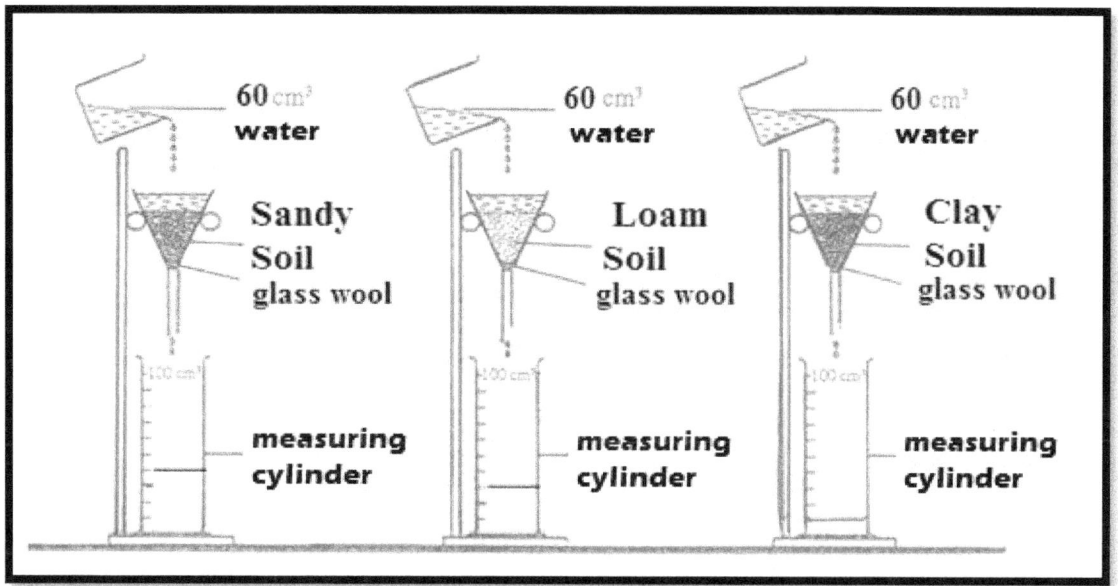

The diagrams show the permeability of different types of soils

The amount of water collected after drainage will show the permeability of the various soil types. The sandy soil having larger particles and air spaces allowed more water to drain. And clay soil with smallest particles and air spaces allowed least water to drain. Loam soil allowed a moderate amount of water to pass through.

Questions

1. As it relates to soils with larger particles, explain the affect it will have on the porosity of soil.
2. What is soil permeability?
3. What determines the soil permeability?
4. Where would most water drain, from soils with larger spaces or soils with small spaces? Why?

Soil Formation

Weathering

Parent or bed rock is the layer of rock beneath the soil. Weathering is the breaking down of rocks and other materials on the Earth's surface. Soil originates from the parent or bedrock through the process of weathering. There are two types of weathering: mechanical and chemical.

Mechanical Weathering

The rocks are broken into smaller pieces and different shapes. At the beginning of process the rocks are sharp and angular. As the process continues, rocks are smooth and have rounded edges.

Mechanical weathering is caused by a number of factors: Temperature, Frost action, Organic activity, Gravity and Abrasion.

a. Temperature

During the day, the sun heats the outside of the rock and it expands. During the night, the outside of the rock cools and contracts. The cycle of heating and cooling continues each day, and parts of rock crack or peel off. This causes exfoliation where the rock breaks off in curved sheets or slabs.

b. Frost Action

Water gets into cracks of rocks, where it freezes. The freezing water expands inside the cracks, and the crack grows until it forces the rock to break.

c. Organic Activity

The roots of plants loosen the rocks. The plant growing in a rock's crack can make the crack grow as the roots grow and spread. This is called root-pry which is the breaking apart of rocks caused by plant roots. Burrowing animals, insects and microorganisms also contribute to soil activity.

d. Gravity

Gravity pulls loose rocks down cliffs of mountains. This is called a landslide. As the rocks fall and collide with other rocks they break them into smaller pieces.

e. Abrasion

Wearing away of solid particles carried in the wind and water. The wind and water pick up particles that have been eroded. The sharp edges of sand and particles cut into the exposed rocks

Chemical Weathering

Weathering that causes changes in the chemical makeup of rocks. Minerals can be added or removed from rocks by chemical weathering. Substances react chemically with rocks and break them down. Some causes are: Water, Oxidation, Carbonation, Sulfuric Acid and Plant Acids.

a. Water

Water can dissolve minerals that hold the rocks together. The water can form acids when it mixes with certain gases. These acids speed up the decomposition of rocks. These acids can combine with mineral to form new substances.

b. Oxidation

This is the process where oxygen combines with another substance to forms a new substance. Example: iron and oxygen combine to form iron oxide (rust). It is shown by its color.

c. Carbonation

Carbonic acid reacts chemically with other substances. Carbonic acid is a weak acid formed when carbon dioxide dissolves in rain.

d. Sulfuric acid

Sulfur oxides are a byproduct of burning coal. When sulfur oxides dissolve in rainwater, they form sulfuric acid. Sulfuric acid is a strong acid that quickly wears away rocks and metals.

e. Plant acids

Plans produce weak acids that dissolve certain minerals. Example: mosses, which grow in damp areas, produce weak acids that seep into rocks and dissolve some minerals breaking down the rocks.

Rate of weathering

The rate of weathering refers to how fast weathering takes place. Weathering is an important part of soil development. It may take place rapidly over a decade or slowly over millions of years. This depends on several factors:

a. Type of Rock - Stable rock can resist chemical weathering longer.

b. Time –the longer a parent material has been exposed, the greater the degree of weathering and the more developed the soil.

c. Size of exposed surface area will determine the speed at which weathering takes place.

Soils and Drainage

Soils with high drainage rates, e.g. sandy soils are not suitable for agriculture because leaching occurs as water passes through quickly carrying valuable nutrients with it. Clay soils retain more water, have poor drainage and easily become water-logged or swampy.

Heating to 100 °C causes water to evaporate from soil sample. The difference in weight of a soil sample before heating and after cooling in a dry environment is equal to the weight of water.

Soil pH

Level Soil pH is an indication of relative acidity or alkalinity. It is reported on a scale of 0 to 14, with low pH numbers being acid and high pH numbers alkaline. A pH of 7 is neutral. In other words, soils can be classified as acidic, basic or neutral. The pH of a soil affects the plants which grow in it. The pH of a soil impacts agriculture. The pH of Bahamian soil is generally slightly alkaline because of the limestone content.

Soil Profile

A soil profile is the arrangement of soil layers from top to bottom. The parent rock is at the bottom with large rocks, smaller ones and soil in successive layers upwards according to size. The order from the top is humus, small clay particles, large clay particles, silt, sand, gravel.

The Soil Profile (Layers)

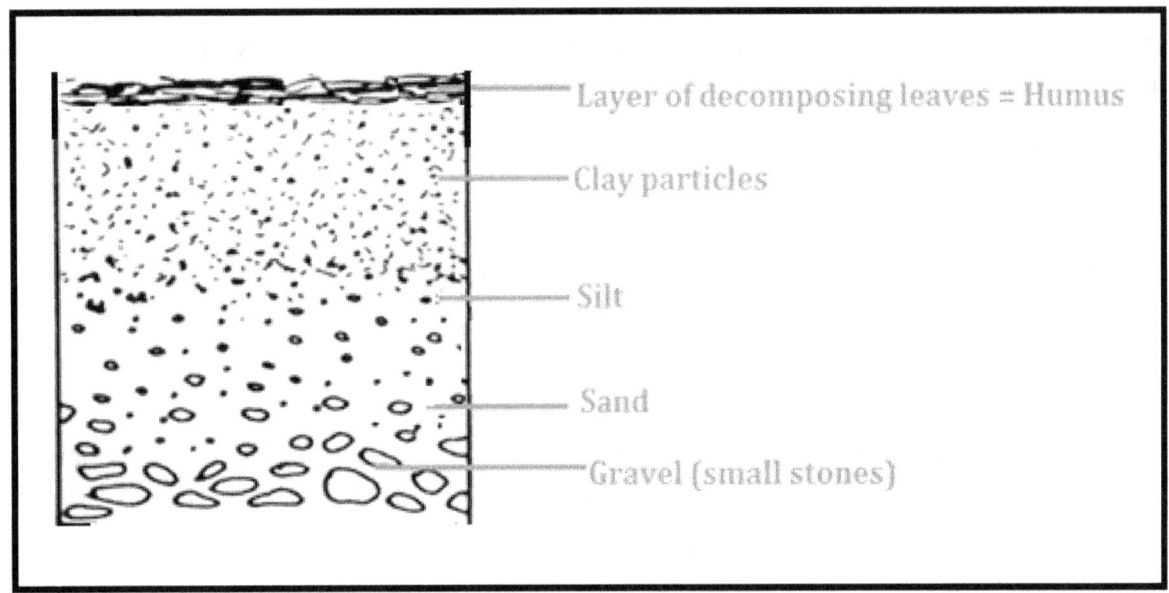

Soil profile

Sedimentation is the process by which matter, e.g. soil settles at the bottom of a liquid, e.g. water. The soil components separate according to their densities. The components with highest densities settle at the bottom and those with the lowest densities settle at the top.

Relationship between Fertility of Soil and Colour

Black coloured soil like loam soil is more fertile than light coloured soils like sandy and clay soils. Black coloured soils contain humus made up of partially or completely decayed plant and animal matter. Humus increases the fertility of the soil. Limestone (calcium carbonate) is the parent rock in The Bahamas, it is essential to white infertile soil.

Unwanted organic matter refers to kitchen waste, garden cuttings, sea weed which is mixed with layers of soil to be decomposed by bacteria in soil.

Types of Soils Based on Colour

Clay soil – white or reddish/brown
Sandy soil- white or yellow/brown
Loam soil – contains humus which adds the black colour to the soil.

Colour depends on mineral content e.g. pink, black.

Comprehension Questions

1. Why are sandy soils not suitable for agriculture?
2. What is the level soil pH? Explain why the pH of Bahamian soils is considered to be alkaline.
3. Define Sedimentation. Where do the lowest densities settle?
4. Explain the relationship between fertility of soil and color.
5. What is the common name for Calcium Carbonate?
6. Create a table differentiating the colors of Clay, Sandy and Loam Soils.

Organic Vs. Inorganic Fertilizers

Fertilizers are chemical substances that release minerals into the soil to become fertile.

Types of Fertilizers

Fertilizers may be natural also known as organic or artificial, known as inorganic.

Organic Natural Fertilizers: Manure and compost release nutrients/minerals slowly and tend to cause slightly acidic conditions.

a. Manure is animal waste. People sometimes spread manure alone on their garden plots. It can be very effective in increasing crop production. Horse, cow, sheep and poultry manure are commonly used this way. Straight manure is high in nitrogen, a main ingredient in commercial fertilizer.

b. Compost can be made up of many ingredients. These include: grass clippings, dead plants, mulched leaves, twigs and other yard waste; kitchen scraps such as banana peels, apple cores, orange and grapefruit rinds, rotting vegetables, coffee grounds and filters, tea bags; paper waste including newspaper, cardboard egg cartons, toilet paper rolls and shredded office paper.

Artificial Inorganic Fertilizers: These are substances that are made in factories that contain the three elements, **nitrogen** (N), **phosphorus** (P) and **potassium** (K) that are most essential for plant growth. Fertilizers containing these elements are called NPK **fertilizers.** They work quickly and are often washed out of the soil by rain. Examples are: Ammonium nitrate, potassium phosphate and potassium sulphate. These fertilizers release minerals relatively quickly.

Soil Erosion

Erosion is the removal of topsoil by the action of wind, rain, deforestation, burning, etc.

Causes of Erosion

Rain- When a raindrop hits soil that is not protected by a cover of vegetation and where there are no roots to bind the soil, it has the impact of a bullet. Soil particles are loosened washed down the slope of the land and either end up in the valley or are washed away out to sea by streams and rivers. Erosion removes the topsoil first. Once this nutrient-rich layer is gone, few plants will grow in the soil again. Without soil and plants the land becomes desert-like and unable to support life.

Deforestation- When the ground surface is stripped of vegetation, the upper soils is vulnerable to both wind and water erosion. Soil is washed into rivers when it rains, and then out to sea. This destroys the ability for the land to regenerate because it has lost its topsoil. It also destroys marine environments.

Overgrazing- Overgrazing occurs when a grazing animal eats too much of the grass or other plant life in an ecosystem. This can cause severe damage to the ecosystem, including the disruption of local food webs and increased soil erosion. Domestic animals like sheep, goats, cows, mules and horses have been responsible for large scale ruin of the vegetative cover.

Wind - moves the air which causes friction against the soil so it slowly erodes away. Wind erosion damages land and natural vegetation by removing soil from one place and depositing it in another. It causes soil loss, dryness and deterioration of soil structure, nutrient and productivity losses and air pollution. Suspended dust and dirt are inevitably deposited over everything.

Soil Conservation

- Replant trees that have been cut down
- Rotate Crops: grow different crop plants at different times through the year. This prevents the soil from being eroded because it is left exposed, prevents soil nutrient depletion and build up of pest.
- Plant leguminous plants e.g. peas, beans, clover. They have nitrogen fixing bacteria in their root nodules that add nitrates to the soil.
- Slough the hillside with the contours and terraces of the hill instead of up and down to prevent the soil from being washed away.
- Use organic (natural) fertilizers, e.g. manure, peat, compost instead of inorganic fertilizers.

Comprehension Questions

1. What is a fertilizer?
2. List the four (4) types fertilizers.
3. What substances make up the NPK Fertilizer? Where are they made?
4. What is the term used to describe animal waste?
5. List five ingredients that can be used to make compost.
6. What is Soil Erosion? What are the causes of soil erosion?
7. Explain one cause of erosion.
8. What are some ways that soils can be conserved?

Activities

1. Collect a sample of soil. Hold lamp over soil sample. Animals move away from light and into container. Transfer the soil to a sieve and shake it until the soil passes through the sieve. Identify the animals that remain in the sieve.

2. Separate different size soil particles using sieves of different pore sizes using larger pores first. Use sieves with a range of mesh sizes.

3. Design and conduct investigations to compare soil porosity. Note the time when water first drips from each soil into the measuring cylinder.

4. Design and conduct investigations to compare drainage in different soils. After water has finished dripping from each soil, record the amount of water in each measuring cylinder.

a. Use measuring cylinders to measure equal volumes of the soils.
b. Use a measuring cylinder to measure equal volumes of water.
c. Measure and record the time it takes for the water to begin to drain from each soil sample.
d. Measure the time it takes for water to stop draining from soil samples.
e. Calculate drainage rate stating the unit, e.g. cm3 of water /s.
f. Draw a conclusion on the suitability of soils investigated for farming, based on its drainage rate determined from the results of experiments performed.

g. Write a statement to compare porosity of the soils investigated, with the rate of drainage and leaching.

5. Make a model of a soil profile.

5. Design and conduct an experiment to show sedimentation. Use a clear, transparent (glass) container/measuring cylinder.

a. Place the soil with mixed particle sizes to about 8 cm high. Use a metric ruler.

b. Add water to an additional 8 cm high.

c. Shake well and allow particles to settle.

d. Observe the appearance of soil samples.

6. Observe air displaced from a soil sample

a. Place 100 ml soil in a 250 ml measuring cylinder

b. Slowly add 150 ml water

c. Observe air bubbles in the water

d. Record final volume of water in the measuring cylinder

7. How to Measure the pH of Soil

a. Half fill a test tube with the soil.

b. Add water and shake well.

c. Test the liquid with a Universal Indicator paper.

d. Record the color developed and determine the pH by comparing the colour with the corresponding colour on the colour chart.

8. Design and conduct an experiment to show sedimentation

a. Use a clear, transparent (glass) container/measuring cylinder.

b. Place soil with mixed particle sizes to about 8 cm high. Add water to an additional 8 cm high.

c. Shake well and allow particles to settle.

d. Observe the appearance of soil samples.

e. Design and conduct investigations to compare the humus composition in soils.

f. In experiments performed to determine the composition of soil, humus floats since it is less dense than water.

The Digestive System

Food is of no use to the body if it is not digested. Food, before it is digested is insoluble. This means it is incapable of being dissolved in a liquid. **Digestion** is the breakdown of insoluble substances into soluble substances that the body can use. The act of taking in food into the mouth is called **ingestion**.

Purpose of Mechanical Digestion

Mechanical digestion is the physical break down of food into smaller pieces. Food must be crushed into smaller pieces and churned into a liquid in order to flow through the alimentary canal easily. One example of this is mastication (chewing) which occurs in the mouth by the teeth, which act to cut and grind food into smaller pieces. This makes them easier to later digest as it increases the surface area of the food molecules.

Parts of the Digestive System

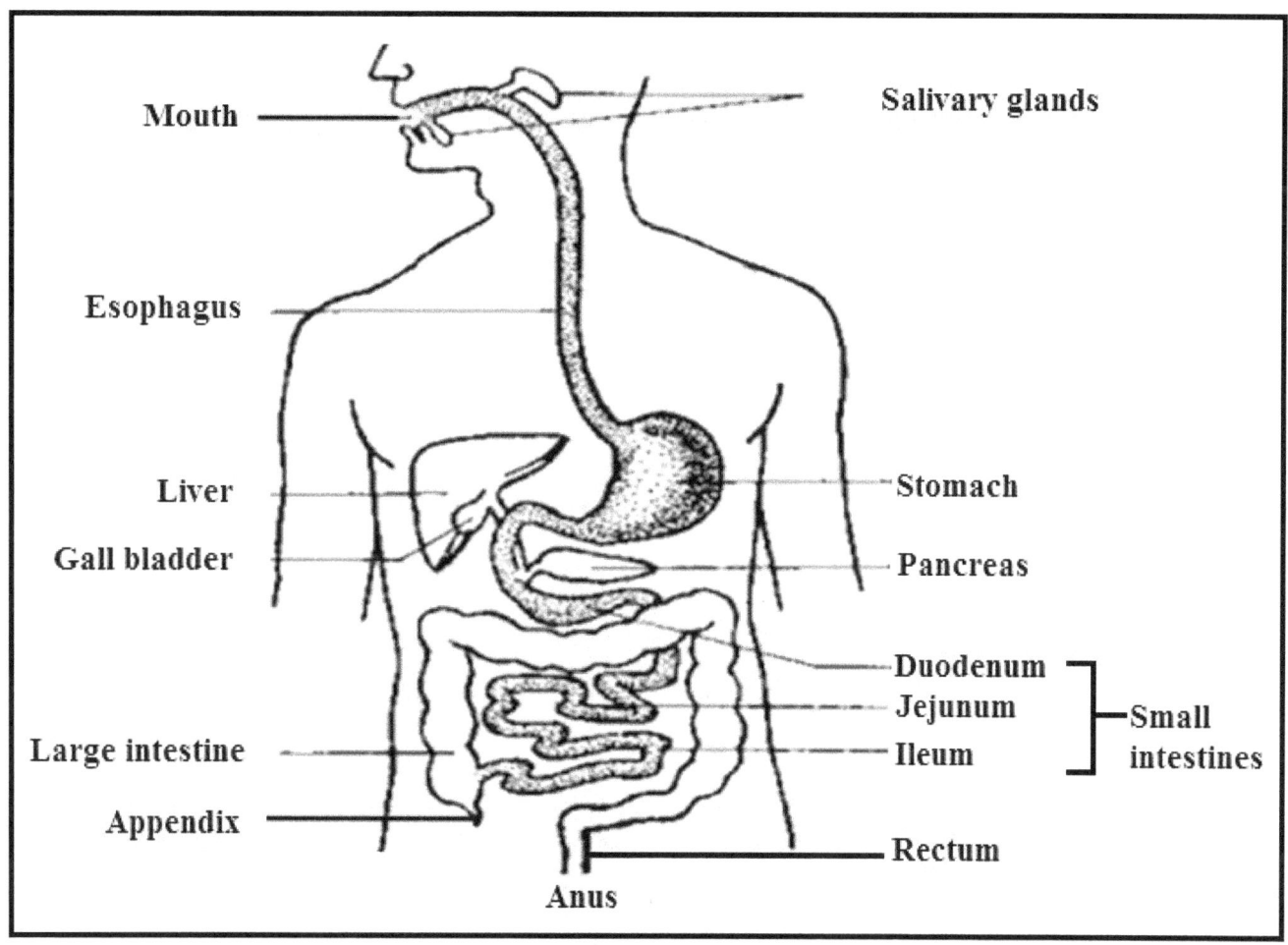

The diagram shows the Digestive System and the accessory organs

The alimentary canal is a tube leading from the mouth to the anus. The digestive system is made up of the alimentary canal and accessory glands which secrete juices into it. Digestion begins in the mouth and ends in the small intestine. The small intestine is also where most absorption takes place.

Mouth

Ingestion is the intake of food through the mouth. In the mouth, the food is chewed and mixed with **saliva**. Food is softened and made smaller in size during the process of mastication. The food can then be swallowed and it also increases the surface area for the enzymes to work on later. Saliva is a **digestive juice** produced by **salivary glands** in the mouth. Saliva contains the enzyme, **salivary amylase.** It is sometimes called **ptyalin**, which acts on cooked starch and begins to break it down into maltose.

The tongue helps to roll the food into a ball called the **bolus** for swallowing. A bolus can be defined as the mass/ball of food that has been chewed and swallowed

The **epiglottis** is a flap of cartilage which closes over the trachea to prevent the food from going down the wind pipe instead of the esophagus. A wave of contraction called **peristalsis** takes place which moves the food through the esophagus, across the **esophageal sphincter** and into the stomach. The esophageal sphincter is a muscle to control the movement of food from the esophagus to the stomach.

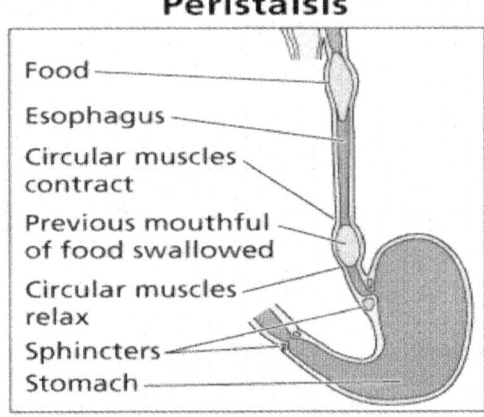

Peristalsis

The diagram shows a bolus through the esophagus by peristalsis

Stomach

As the food enters the stomach from the esophagus, the stomach walls stretch because of its elasticity. The **pyloric sphincter** is a band of muscles at the lower end of the stomach which stops solid pieces of food from passing through into the small intestine prematurely. The main functions of the stomach are:

- to store the food from a meal,

- turn it into a liquid
- release food in small quantities at a time to the rest of the alimentary canal.

Glands in the lining of the stomach produce **gastric juice** containing the enzyme pepsin. **Pepsin** is a protease, it acts on **proteins** and breaks them down into soluble peptides. **Rennin** in the stomach clots milk and is important in infancy. Very little absorption takes place in the stomach. However, some drugs like aspirin and alcohol take effect very quickly because they are partly absorbed by the stomach walls.

Hydrochloric acid is also made in the stomach. This acid provides the best degree of acidity for pepsin to work and it kills many of the bacteria taken in with food. The stomach wall is therefore protected by mucus.

The pyloric sphincter lets the liquid products of digestion pass, a little at a time, into the first part of the small intestine called the duodenum.

Small Intestine

The small intestine is roughly 6 meters long, the large is 1.5 meters. The diameter, not the length, differentiates the small and large intestines.

The small intestine is the region in which the greater part of digestion takes place. This is also where digestion is completed and absorption takes place. Absorption is the movement of digested food from the intestines into the blood stream. The small intestine is made up of three segments:

- **Duodenum:** This short section is the part of the small intestine that takes in semi-digested food from the stomach through the pylorus, and continues the digestion process. The duodenum also uses bile from the gallbladder, pancreatic juice from the pancreas and intestinal juice from the intestinal glands to aid in the digestive process.
- **Jejunum:** The middle section of the small intestine carries food through rapidly, with wave-like muscle contractions, towards the ileum.
- **Ileum:** This last section is the longest part of your small intestine. The ileum is where **most of the nutrients from food are absorbed** before emptying into the large intestine.

The Final Products of Digestion

When digestion is completed, the digestible materials are those that can pass through the lining of the intestine and into the blood stream are:

Food		Final Product
Starch	$\longrightarrow$	glucose
Fats	$\longrightarrow$	fatty acid and glycerol
Proteins	$\longrightarrow$	amino acids

The small intestine carries out most of the digestive process, **absorbing** almost all of the nutrients you get from foods into your bloodstream. The walls of the small intestine make digestive juices, or enzymes, that work together with enzymes from the liver and pancreas to do this.

The small intestine has a number of features which allow the food to be taken up by the blood, a term called **absorption.** Villi (villus: singular) are found in the walls of the small intestine. They are fingerlike projections with large surface area. Increased surface area and the time food spends passing through the small intestine, so increasing chances of absorption. Each villus has many microvilli.

The Villi

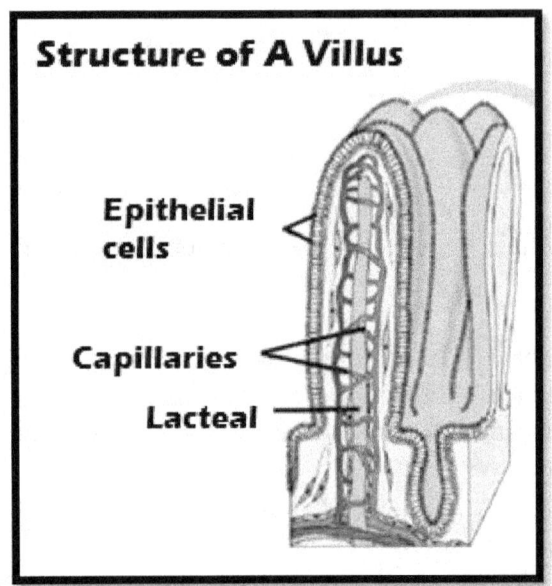
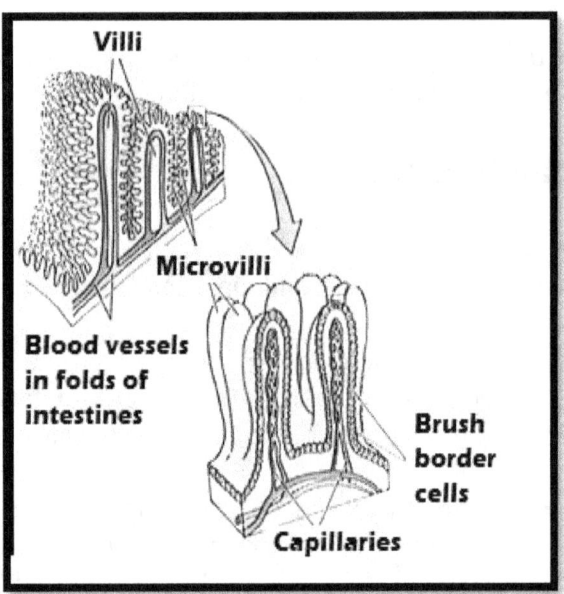

The diagrams show the structure of the villi   The diagram shows a the position of microvilli

A villus is covered by types of cells known as epithelium. It is the epithelial cells that actually help absorb, move, and distribute some of the fluids and nutrients in the body. A villus also has a network of **capillaries**, which are very tiny blood vessels. These blood capillaries absorb digested **proteins in the form of amino acids and carbohydrates in the form of glucose**. The villus has another type of tiny vessel known as a **lacteal**. Lacteals absorb digested fats and oils in the form of fatty acids and glycerol.

Features of the Villi

- It is fairly long and presents a large surface area.
- The lining epithelium is very thin and the fluids can pass rapidly through it.
- There is a dense network of blood capillaries.

- **Mucosal folds:** The inner surface of the small intestine is not flat, but thrown into circular folds.
- **Villi:** The folds form numerous tiny projections which stick out into the open space inside your small intestine (or lumen), and are covered with cells that help absorb nutrients from the food that passes through.
- **Microvilli:** The cells on the villi are packed full of tiny hair-like structures called microvilli. This helps to increase the surface of each individual cell, meaning that each cell can absorb more nutrients.

After the digested food has been absorbed into the blood stream, it is transported by the blood to the cells and tissues of the body. The process of digested food entering the body cells and tissues is called **assimilation.**

Where Digestion Begins and Ends for Each Class of Nutrient

Carbohydrates – cooked starch begins in mouth, ends in ileum.
Proteins - begin in stomach ends in ileum.
Fats – begin in duodenum, ends in ileum.
Vitamins, minerals and water are readily taken up into the bloodstream and do not have to be digested.

Enzymes are chemicals that speed up the rate of chemical digestion significantly. Enzymes are biological catalysts.

Digestive juices aid in the process of chemical digestion (e.g. bile, pancreatic juice, trypsin)
Saliva breaks down starch (amylase/ carbohydrase)
Gastric juice breaks down on protein (protease)
Pancreatic juice breaks down starch (amylase/ carbohydrase), protein (protease), fats (lipases)
Intestinal juice breaks down starch (amylase/ carbohydrase), protein (protease), fats (lipases)

Large Intestine

The large intestine is much broader than the small intestine and takes a much straighter path through the abdomen. Its job is to **absorb water and salts** from the materials that have not been digested as food, and get rid of any waste products left over. By the time undigested food reaches your large intestine, what is left is mainly fiber, dead cells, salt, bile pigments (which give this digested matter its color), and water. In the large intestine, bacteria feed on this mixture. The large intestine is made up of the following parts:

- **Rectum:** The final section of digestive tract. Leftover waste collects in the rectum, expanding it until you go to the bathroom. At that time, it is ready to be emptied through your anus, a term called **egestion.**

Accessory Organs

Some organs are a part of the digestive system, but food does not pass through these organs. They are called accessory organs because they aid in the digestive process. The salivary glands, the liver, gallbladder and pancreas are all accessory organs. They produce or store chemicals which aid in the digestion of food. These organs and their functions are listed on the chart below.

Organs, Digestive juices, Enzymes and Actions

Organ	Region	Digestive Juice	Name of Enzyme	Action / Function
Salivary Glands	Mouth	Saliva	Salivary amylase	Begins the digestion of starch to produce maltose
Stomach	Stomach	Gastric juice	Pepsin	Turns proteins to peptides
Gall bladder	Small intestine	Bile	Not an enzyme	Turns fats into fat droplets
Pancreas	Duodenum	Pancreatic juice	Amylase Lipase Proteases	Helps to digest starch to maltose protein to peptides and amino acids and fat into fatty acids and glycerol

Practices to Maintain a Healthy Digestive System
- Eat a balanced diet that includes all nutrients, much water and fiber, which promote regular bowel movements
- eat on time
- avoid late night meals

Disorders of the Digestive System

Blockage (Growth of Tissue) in the Small Intestine

Food travels the alimentary canal as peristaltic contractions. Obstructions sometimes referred to as blockage, prevent the normal flow of food and fluid. When this happens, there is a back-up of food causing pain and sensation of being "full". Reduced defeacation, adhesions, hernias or tumors are all causes of intestinal obstruction or blockage.

Symptoms

Signs and symptoms of intestinal obstruction include:

- Crampy abdominal pain that comes and goes

- Nausea
- Vomiting
- Diarrhea
- Constipation
- Inability to have a bowel movement or pass gas
- Swelling of the abdomen (distention)

Causes

Mechanical obstruction of the small intestine
Common causes of mechanical obstruction, in which something physically blocks the small intestine, include:

- **Intestinal adhesions** — bands of fibrous tissue in the abdominal cavity that can form after abdominal or pelvic surgery
- **Hernias** — portions of intestine that protrude into another part of your body
- **Tumors** in the small intestine
- **Inflammatory bowel diseases**, such as Crohn's disease
- **Twisting of the intestine** (volvulus)

Diet for a Person Whose Gall Bladder was Removed
Gall bladder stores bile which aids in the digestion of fatty foods. Diet should have minimal fatty component. Bile emulsifies (breaks up) fat into small "droplets" so increasing the surface area for enzymes to work. Bile is made in the liver and stored in the gall bladder.

A person whose gall bladder was removed would sometimes have difficulty in handling fats in the diet. Fat and certain fat-soluble vitamins require bile in order to be absorbed. When the gallbladder is present, it stores bile that the liver makes. During a meal, the gallbladder contracts, releasing a pool of bile into the intestine that is used for fat absorption. After the gall bladder is removed (cholecystectomy), bile is still produced by the liver, but is released in a continuous, slow trickle into the intestine. When eating a meal that is high in fat content, there may not be enough bile in the intestine at the time to properly handle the normal absorption process. The change in intestinal bile concentration during high-fat intake may cause diarrhea or bloating, because excess fat in the intestine will draw more water into the intestine, and because bacteria digest the fat and produce gas. Some studies suggest that diarrhea after cholecystectomy may also be caused by excess bile in the intestine between meals, because bile is released into the intestine continuously.

Constipation is the inability to pass faeces.

Diarrhoea is the continual defecation of faeces usually in a liquid form.

Gastroenteritis is inflammation of the intestines.

Liver and Pancreas in Digestion and the Endocrine System

In Digestion
Liver

- makes bile salts, emulsify fats
- stores converts glucose to glycogen
- breaks down excess amino acids

Pancreas
- secretes pancreatic juice
- protease
- lipase
- amylase

In The Endocrine System
Liver
- controls amount of sugar in blood
Pancreas
- makes hormones (chemicals) to monitor sugar

Tripe
"Tripe" is the common name referring to the long, narrow, white intestine of sheep, cows, pigs. The inner wall lining is rough with "bumps".

Green tripe is simply any tripe that has not been washed and bleached. It is used as dog food and is usually made from the fourth stomach of a cow because the texture of that stomach compartment is not suitable for human consumption.

"When tripe is harvested and prepared for humans, the contents of the stomach are removed and discarded, and the lining is washed thoroughly. At this point, the fresh clean tripe has a characteristic khaki color which most people find rather unappealing. Therefore the tissue is bleached to yield the snow-white product we find in our grocer's meat department.

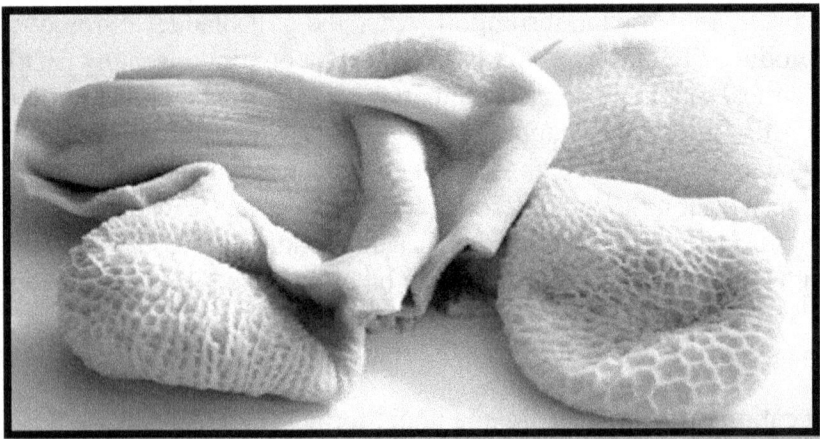

Beef Tripe

Activity:
Obtain a sample of tripe from a Butcher and observe the outward appearance and use a hand lens to observe the surface of the inner lining. With the aid of diagrams, describe the features of "tripe."

Comprehension Questions

1. Draw and label the parts of the digestive system.
2. Why is it necessary for food to be digested?
3. State where digestion:

    a. begins
    b. ends

4. List the parts of the digestive system in order. Do not include the anus.
5. Tell the functions of
        a. bile
        b. the gall bladder
        c. the liver
        d. and pancreas

6. Describe how food is moved through the alimentary canal.

7. What prevents food from going into the trachea?
8. Define the term bolus. Use a diagram to help you.
9. Name TWO effects of hydrochloric acid in the stomach.
10. What is the function of digestive enzymes?
11. State briefly what happens to protein in food from the time it is swallowed, to the time its products are built up into the cytoplasm of a muscle cell.

12. Use the Venn Diagram to compare and contrast the pancreas and liver.

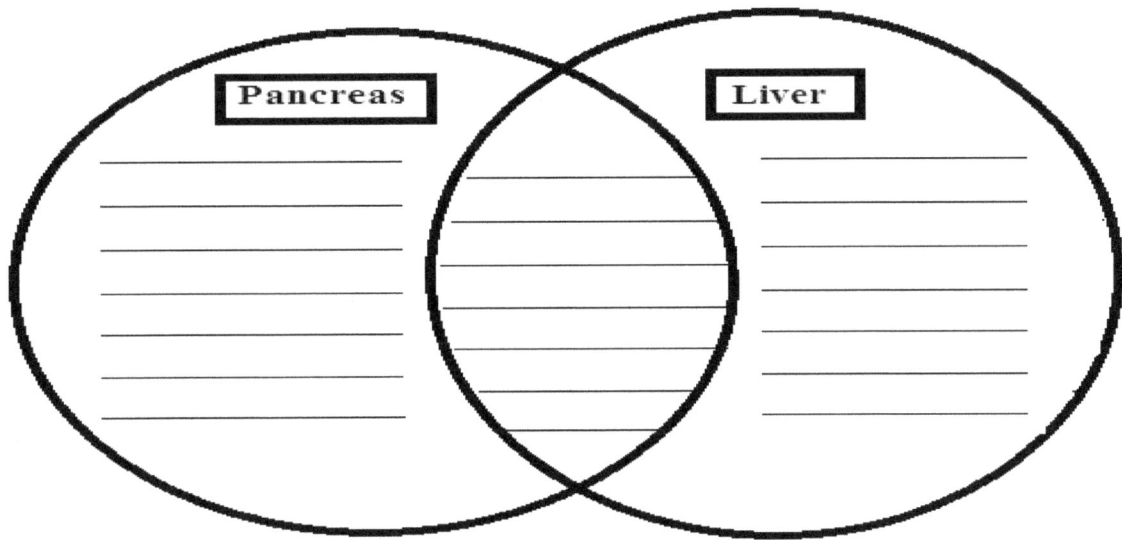

13. Draw and label the parts of a villus.
14. Your teacher will give you a diagram of the digestive system. On the diagram, colour the organs in which digestion begins and ends for:

a. Starch   Red
b. Proteins   Orange
c. Fats / Oils   Purple

15. What are the products of digestion?

a. Starch
b. Protein
c. Fats, which are absorbed in the ileum?

16. List THREE signs of intestinal obstruction
17. Compile a list of foods that people without gall-bladders should avoid (due to inadequate amounts of bile to process them at mealtime).
18. Indicate the role of each in the digestion process
19. Define the following terms:

a. Ingestion

b. Digestion

c. Absorption

d. Assimilation

e. Egestion

Activities

1. Colour the parts of the digestive system.
2. Name the individual parts or organs.
3. Make a (life-sized) model of the human digestive system using various materials.
4. Identify the villus / villi.
5. Draw the villi and cross section of the intestinal walls.
6. View a villus using a microscope.
7. Measure a string to length of small intestine. Measure a different colour string / yarn to the length of the large intestines.
8. Compare lengths of the intestines.
9. List mechanical changes and chemical changes foods undergo.
10. Simulate mechanical changes of food.
11. Move a paper ball along a tube or flexible cylinder. Simulate muscular motions.
12. Class discussion -Identify and adopt practices to maintain a healthy digestive system.
13. Use starchy foods to investigate the rate at which enzymes act on substrate.
14. Identify and adopt practices to maintain a healthy digestive system.

Technology

The word **technology** refers to the making, modification, usage, and knowledge of tools, machines, techniques, crafts, systems, and methods of organization, in order to solve a problem, improve a preexisting solution to a problem, achieve a goal, handle an applied input/output relation or perform a specific function. It can also refer to the collection of such tools, including machinery, modifications, arrangements and procedures. Technologies significantly affect human as well as other animal species' ability to control and adapt to their natural environments. The term can either be applied generally or to specific areas: examples include *construction technology*, *medical technology*, and *information technology*. Information technology will now be discussed as it relates to communication.

Communication through technology is a continually evolving process from the light houses to the modern day forms of communication systems.

Lighthouses

A lighthouse is a tower designed to emit light from a system of lamps and lenses (or, in older times, from a fire) and used as an aid to navigation and to pilots at sea. The lighthouses were a means of communicating to pilots at sea. The lighthouses warn sailors away from rocks, reefs and shallow waters. They also let you know where you are. As you travel along the coastline, you need some landmarks to help you find your position. To help the mariner determine his location, each lighthouse was painted in different colors and designs. The pattern or color of a lighthouse is called its **day mark.** When you know the day mark of each tower, you always know where you are during the day.

You can't see colors or patterns at night, but you can see lights. However, unless there was some way to make each light different, you could have the same problem. A light can send out a flash every five seconds, or it might have a fifteen second period of darkness and a three second period of brightness, or any number of other combinations. The individual flashing pattern of each light is called its CHARACTERISTIC. Mariners have to look at a light list or a maritime chart which tells what light flashes that particular pattern and what color the light is as well. Then they are able to determine their position at sea in relation to the land. You might find the lighthouse standing alone, attached to the building where the lighthouse keeper lives, or connected to the keeper's quarters by an enclosed walkway. This is called a light station. Sometimes the lantern room is built into the roof of the keeper's house.

Examples of Lighthouses with Day-Markers

The pictures show lighthouses with various day markers

A lighthouse is usually tall, cylindrical in shape, wide at base and narrower towards the top. The three sections are: **main (service), optic (light) and dome (cover).**

Sections of a Lighthouse

The **service room** is where the keeper cleaned the lamp chimneys and prepared the lantern for the coming night.

The **optic section** is the area surrounded by glass windows called storm panes. The storm panes are set in metal frames. A lighthouse optic is an assembly of lenses and prisms. It is used to gather as much light from the light source and send it out of the lighthouse in a beam with a characteristic flashing sequence. For many, many years the "optic" section was open to the elements (no glass storm panes) and was the area where an open fire burned. A Fresnel (pronounced "Frey Nel") Lens is a type of optic consisting of a convex lens and many prisms of glass which focus and intensify the light.

Optic Section   Fresnel Lens

The **dome** is made of metal and usually copper clad. It is surmounted by a ball vent through which air can pass. The ball vent is topped by a lightening rod and sometimes a wind vane to assist the keeper in discerning wind direction which will help him position the vents. In the interior of the dome, beneath the top, is a large concave dish. This "dish" catches condensation that forms in the top of the lantern.

Existing Lighthouses in the Bahamas
Lighthouses are found on the islands of the Bahamas listed below:

Abaco (South tip)	Hole in the wall
Abaco	Hopetown
Berry Islands	Great Stirrup
San Salvador,	Dixon Hill Lighthouse
Cay Lobos,	Off the North Coast of Cuba, but it is owned by the Bahamas.

Lighthouses in the Bahamas

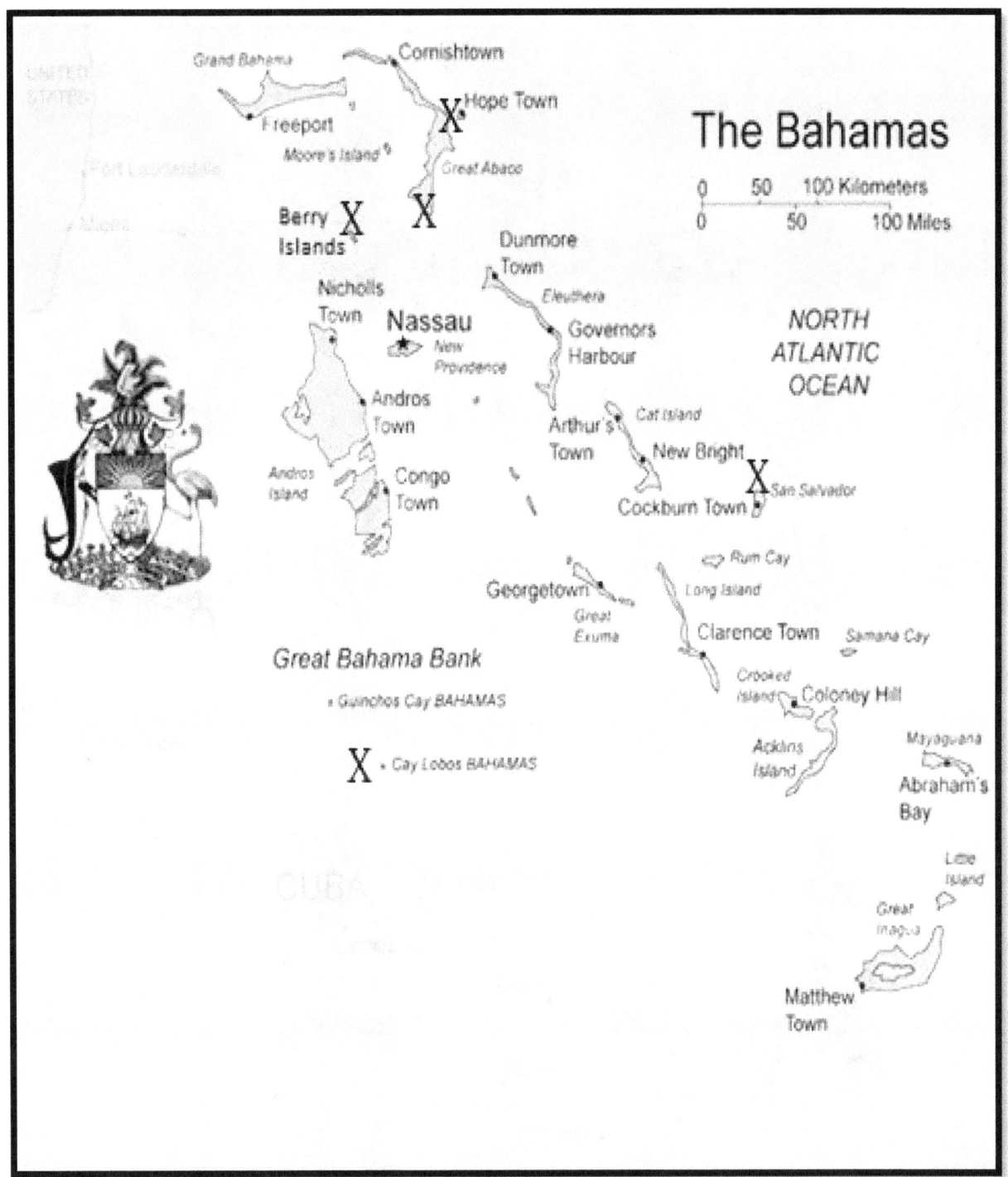

Map of the Bahamas

The letter "X" indicates the island where a lighthouse may be found

Parts of a Lighthouse

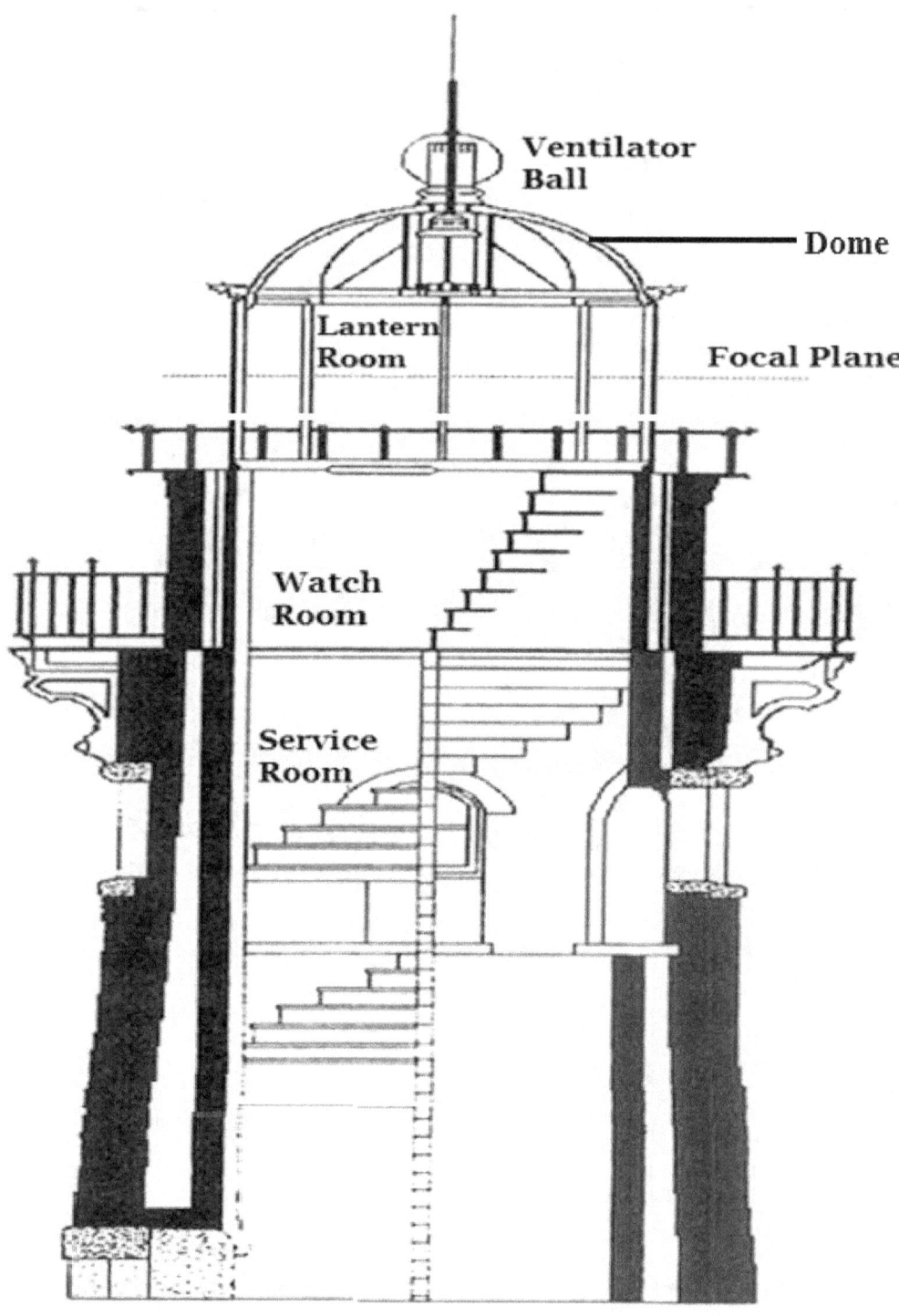

The diagram shows the main parts of a Lighthouse

Lighthouses and the Presence of Reefs

The Bahamas consists of several hundred islands, cays, and rocks; many of which are submerged (reefs). Ships become wrecked on reefs at night because the pilots are unable to see. The lights from lighthouses warn ships of the presence of reefs.

Brightness of Lights

Lights must be seen over great distances and in bad weather. Some lights are 500 and others 1,000 watts.

Types of Lights and Effectiveness

Incandescent kerosene lamp with a revolving lens which magnified the brightness of the light was commonly used in light houses. However, electrically powered lights, electronically remote-controlled and solar powered lights have replaced kerosene over the years.

Different fuels were used to illuminate the lamps over time. First, it was wood and coal for fires, then bales of oakum and pitch, and rows of candles. Later lamps were lit using various fuels - sperm whale oil, lard oil from animal fat, kerosene, etc. When the first lighthouse was designed with an enclosed lantern room it was possible to use candles for light. One lighthouse used 60 candles! Most used far fewer candles which were sometimes arranged in a circular candelabra or a chandelier with two tiers, or on a frame.

Scientific Principle for Flashing Lights

As the lens rotate around the light it produces the flashing". Most lenses contain hundreds of prisms which are pieces of specially ground, cut and polished glass. When they are arranged in a certain way, they bend (reflect and refract) the light. Thus, all the rays of the light are collected and redirected into a single pencil beam of light. This makes it much brighter and more effective. Frequency of flashes identifies the lighthouse.

Factor That Might Have Limited the Effectiveness of Lighthouses

Weather conditions: Most lighthouses were operated manually (depended on a person). Storms and fog were factors that might have limited the effectiveness of lighthouses. If the view of the light from the lighthouse was blocked by fog for example, another method of notifying the mariner was used. This method was to use sound. It was called a foghorn similar to cannon. The lighthouse keeper would have to fire the foghorn every hour during a long spell of fog. Because of this, the keeper would not get any sleep. Later they tried other means of making a noise for warning. Fog bells were used as well as steam whistles, reed trumpets and sirens. The sounds they gave out were generally low pitched and very mournful - almost like a wail. Each one emitted a specific number of blasts every minute so it could be told apart from all others. Today, an automatic sensor which detects moisture in the air turns on the fog signals when needed. There are also soundless fog signals called radio beacons (an electronic device). These fog signals were not placed everywhere. Although some places experience no fog problems, fog warning devices are very necessary in some areas.

Many of the lighthouses are now no longer needed because of advances in technology. The Global Positioning System (GPS) revolutionized maritime navigation when it was first introduced for civilian use in 1994. GPS navigation aboard ships provides the mariner with access to the benefits of real-time positioning data, not only by simplifying the navigation process, but by minimizing penalties for early or late arrivals in port.

Lighthouses in our country have been automated with the use of electronic or solar powered lights etc. Many have been or are being turned over to various government agencies or non-profit local organizations to maintain and administer. It is important to keep them in good condition for future generations to learn about their place in the history of our country. They also need protection from vandalism and threats of erosion. It is a special experience to be able to climb the stairs just as the keepers did and picture what life was like in times past.

Comprehension Questions

1. What is a lighthouse?
2. What do lighthouses look like?
3. How can one lighthouse be distinguished from another?
4. Where are lighthouses located in the Bahamas?
5. What happens during conditions like fog when the light is not visible?
6. What is the difference between a lighthouse and a light station?
7. Do you think light houses are important today? Give at least TWO reasons for your answer.

Activities

1. Observe photographs or visit a lighthouse.
2. Mark and label the location of existing lighthouses on a map of the Bahamas.
3. In small groups, write one verse of a poem / rap or a jingle describing the role of the light house.
4. Use maps to estimate the distance each light must cover.
5. Investigate the effect of lenses (different kinds) on a beam of light (flashlight or similar source).
6. In small groups, make a model lighthouse.
7. Research the scientific principle for the flashing lights.

Matching

Match the definitions with the correct term.

DEFINITION	TERM
1. _____ A room immediately below the lantern room where fuel and other supplies were kept. This is where the keeper prepared the lanterns for the night and often stood watch. The clockworks (for rotating lenses) were also located there.	a. Light station
2_____.A complex containing the lighthouse tower and all of the outbuildings, i.e. the keeper's living quarters, fuel house, boat house, fog-signal building, etc.	b. Refract
3. _____A lighted beacon of major importance in navigation.	c. Fog signal
4. _____To or bend light from a straight path.	d. Light station
5. _____Individual flashing pattern of each light.	e. Lens
6. _____Unique color scheme or design which identifies a specific lighthouse during daylight hours.	f. Fresnel Lens
7. _____Rounded roof on top of some lighthouse towers.	g. Lantern room (Optic section)
8. _____A device (such as a whistle, bell, cannon, horn, siren, etc.) which provides a specific loud noise as an aid to navigation in dense fog.	h. Dome or cupola
9. _____A type of optic consisting of a convex lens and many prisms of glass which focus and intensify the light through reflection and refraction.	i. Fuel
10. _____Material that is burned to produce light (fuels used for lighthouses included wood, lard, whale oil, tallow, kerosene). Today, besides electricity and acetylene gas, solar power is also used.	j. Characteristics
11. _____A curved piece of glass for bringing together or spreading rays of light passing through it.	k. Service room
12. _____ Glassed-in housing at the top of a lighthouse tower, containing the lamp and lens	L. Lighthouse

Basic Principles of Telegraphs

In 1877, Samuel Morse used electricity to make the first telegraph. This invention allowed people to communicate directly with one another over long distances. With telegraphs, a message was sent as electric current through a wire to a device that made dots and dashes to match the message.

The first telegraphs used a simple electromagnet that moved a lever with a stylus into contact with a moving strip of paper. The stylus would leave a long mark or a short mark, depending on how long the operator at the other end held down the telegraph key.

International Morse Code Principles

1. The length of a dot is one unit
2. A dash is three units
3. The space between parts of the same letter is one unit
4. The space between letters is three units
5. The space between words is seven units.

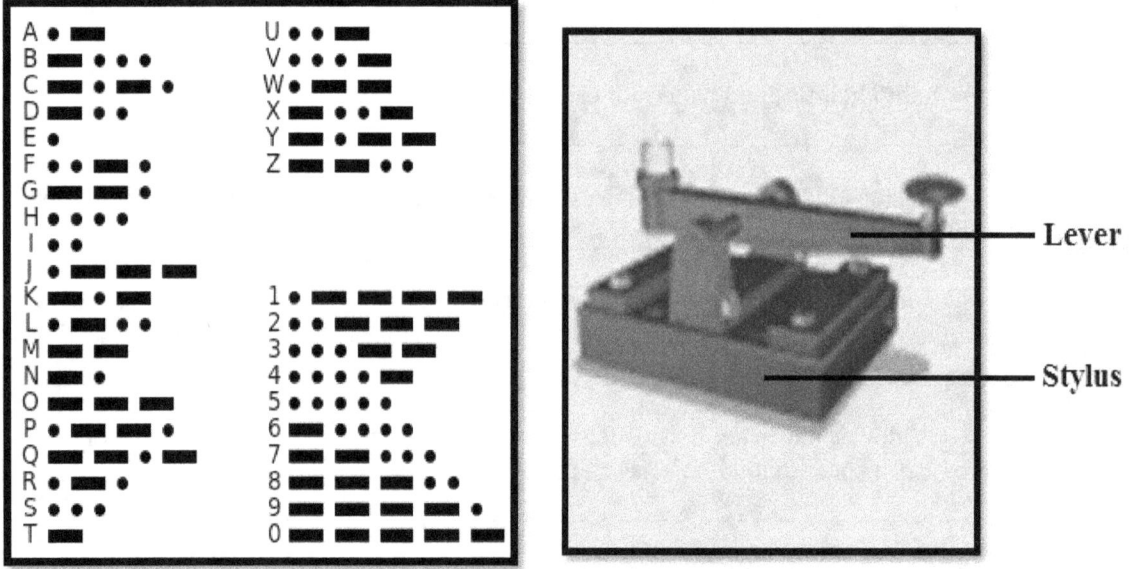

Morse Code   Stylus and Lever

Basic Principle of Telephone

A telephone is an instrument that sends and receives information, usually by means of electricity. The word telephone comes from the Greek words meaning far and sound. The telephone is one of our best ways to communicate. Alexander Graham Bell invented the telephone in Boston in 1876. One hundred and twenty (120) years later there are over 360 million telephone numbers, and that figure grows each year.

The sound from words is changed into electrical signals by the electromagnet in the "mouthpiece". Electrical signals travel through wires and are exchanged to the "ear piece" where the electrical signals are changed to corresponding sounds (words) by the electromagnet in the loudspeaker.

The most familiar telephone is the desk telephone, which sits on a desk, table or shelf. Some phones have options like holding multiple calls or transferring calls to other phones. An intercom allows you to talk to other people in other rooms. Speaker phones have a microphone and a loud speaker. With a speaker phone more than two people can talk in a conversation. Cordless phones do not have wires connected to them, that is why they are called cordless phones, but they still need to have some nearness to a unit that is wired to the telephone system. Cellular phones are true wireless phones.

Components of a Telephone

The part of a telephone that a person picks up to make a phone call is the **handset**. It has an earpiece and a mouthpiece. There is a loudspeaker in the earpiece to amplify what you hear and a microphone in mouthpiece to amplify what you say. The two parts are connected by electrical wires.

The Handset of a Telephone

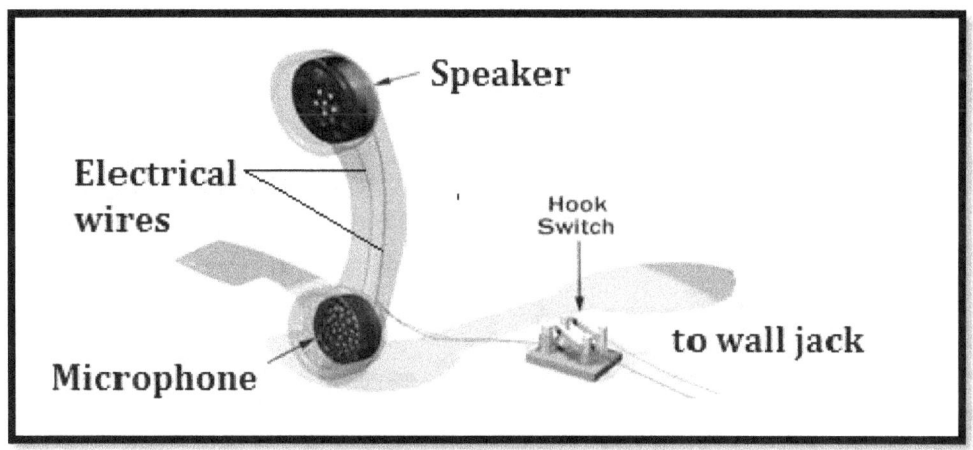

Parts of the handset of a telephone

A telephone has three main parts: (1) **a dialing mechanism**, (2) a **transmitter**, and (3) a **receiver**.

The dialing mechanism enables a caller to enter telephone numbers. The dialing mechanism may be built into the handset, between the earpiece and the mouthpiece. Or it may be part of a separate base unit that is connected by a cord to the handset.

In most telephones manufactured since the 1960's, the dialing mechanism consists of a set of buttons or keys called a keypad. A standard keypad has 12 keys, the digits 0 through 9, a "*" key, and a "#" key. When pressed, each key generates either a certain number of electric pulses or a pair of accurately controlled tones. Computers in the telephone network use the sequence of pulses or tones to direct the call.

The transmitter, also called the microphone, converts the sound waves of a person's voice into an electric current and sends this current farther into the telephone network. The transmitter is built into the handset, behind the mouthpiece.

Over the years, though styles have changed, the basic components have remained the same in telephones.

Telephones used over the Years

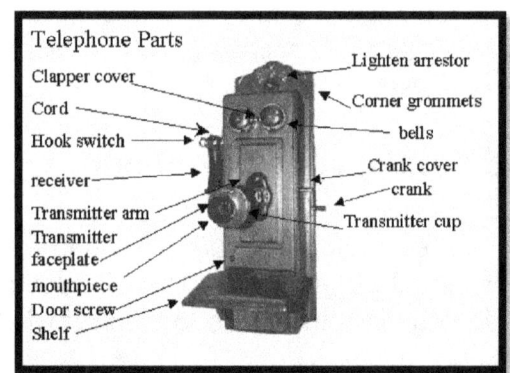

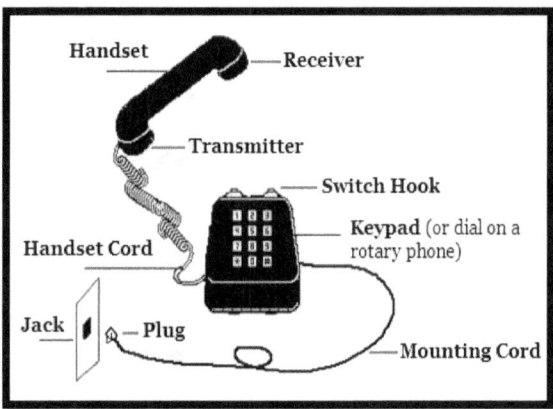

Rotary Phone   Touch Key Pad Phone

Similarities and Differences between Morse Code and Telephone

- Telegraphs generally required skilled operators who knew Morse code, and so most people never had telegraph machines in their homes. In order to send a telegram, one had to go to a local telegraph office. The telephone, on the other hand, requires no special skills, you only needed to speak into the mouthpiece (microphone), and listen through the earpiece. In time, sophisticated switching networks were developed that allowed telephone users to dial each other directly. Similar technology was used for telegraphy (telex) until quite recently.
- The earlier invention, the telegraph, was actually a digital device in the sense that the data it transmitted consisted of simple on-off patterns, and its output was a string of fixed symbols. Now, of course, cellular telephones are used to send text messages as well as voice messages. Telephones and telegraphs are again united.

Activities

1. Observe the components of a telephone.
2. Make a model telephone.
3. Write a simple story (narrative, conversational, cartoon) to describe the operation of the telephone.

The Fax Machine

In today's business environment, the fax or facsimile machine is an integral part of day- to-day operations. In many instances, it has replaced conventional mail as a means by which to convey information. Modern offices commonly send purchase orders, preliminary contracts, employment information, and other documentation in this fashion. The fax machine is used to send a copy of documents to another fax machine without the document actually being sent.

Basic Principle of the Fax Machine

Sending fax machine scans document converting the black markings (letters/lines) and white spaces into electrical signals to be sent to receiving fax machine which changes electrical signals via heat to black markings with white spaces making an identical copy – facsimile.

The machine itself is connected to standard phone lines. A document is inserted into the machine, where it is fed using a rotating cylinder. A light beam scans the document as it passes and the light spaces in the document are translated into pulses of electric current by the photoelectric cell. Dark spaces in the document produce no pulses and grays are interpreted according to the intensity of the tonality.

Parts of the Hp Fax Machine

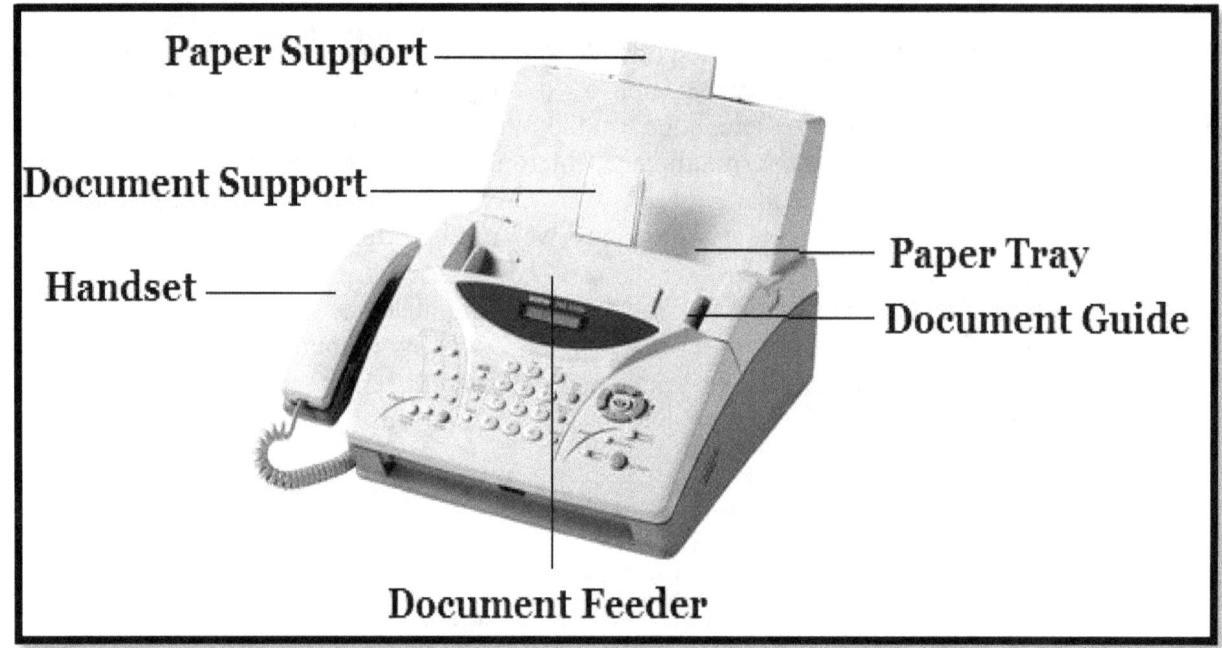

Steps for Using a Fax Machine

1. Check to make certain that its telephone cord is plugged into a phone jack.
2. Insert the document into the "outgoing" fax tray.   Most fax machines require you to insert the documents face down.
3. Dial the phone number to which you wish to send the document.
4. Wait for the receiving fax machine to "answer." Depending on your fax machine, you may have to press the "Start" button to send your fax, or your model may automatically start sending the fax by itself.
5. Look at the fax machine display console. It shows if pages have gone through successfully or if you need to resend a fax due to an error. When a fax has gone through successfully, the machine will beep or display a "success" message.
6. Always ensure that your fax machine is stocked with plenty of paper in the "incoming" fax tray. If you receive a fax, it instantly prints on the paper provided. Even if you are out of paper, your fax machine will keep received faxes in its memory, and it will print them when you finally do stock the machine with paper.
7. If your phone has its own phone line, it should receive faxes automatically. If you use one phone line for both your fax machine and a regular telephone, you may need to press "Start" to process an incoming fax. You can recognize an incoming fax easily; if you pick up the phone, you will hear a similar sound as when you are sending a fax.

Activities

1. Formulate a hypothesis on the scientific principle(s) for the fax machine.
2. Observe components of a fax machine.
3. In pairs, measure the length of the laboratory / classroom.
4. Predict the time taken for sound from a radio to travel the length of the room.

Scientific Principle of the Radio

Radios are used for many purposes. Some examples are communication, radar navigation and television broadcasting. Radios affect everyone's life in many ways. Radios help us get the weather reports, they help NASA speak to astronauts; they even allow us to speak to our friends on the telephone.

Sounds (words/music) go into a microphone which changes them to the appropriate electrical signals which, in turn, are combined with other (carrier) waves. These waves are sent to a tower which "beams" them out as radio waves to the aerial of a radio (receiver). The loudspeaker in the radio changes the radio waves back into the sound waves.

Most radio and television shows are broadcast as radio waves. These are a band of waves in the electromagnetic spectrum with a range of different frequencies and wavelengths. Radio waves are the longest waves in the electromagnetic spectrum.

Before broadcasting, sounds have to be converted into electrical signals by microphones.

Modulation
To enable them to be broadcast, electrical signals have to be altered, using a method called modulation. This is done by mixing the electrical sound signal with radio waves called carrier wave.

As a result of modulation, the shape of the carrier wave varies depending on the electrical sound signals. With **frequency modulation** (FM) the electrical signals are altered to match the frequency of the carrier wave. With **amplitude modulation** (AM) the electrical signals are altered to match the amplitude (strength) of the carrier wave.

Amplitude Modulator

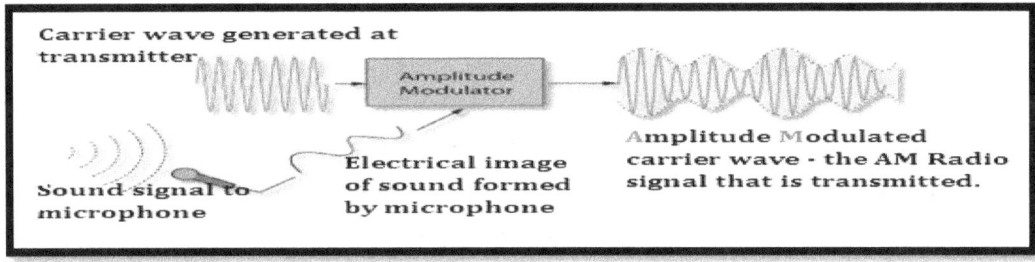

The diagrams show carrier waves and amplitude modulation (AM) carrier wave

A radio works by receiving modulated radio waves through its antenna and then converting them back into very weak electrical signals.

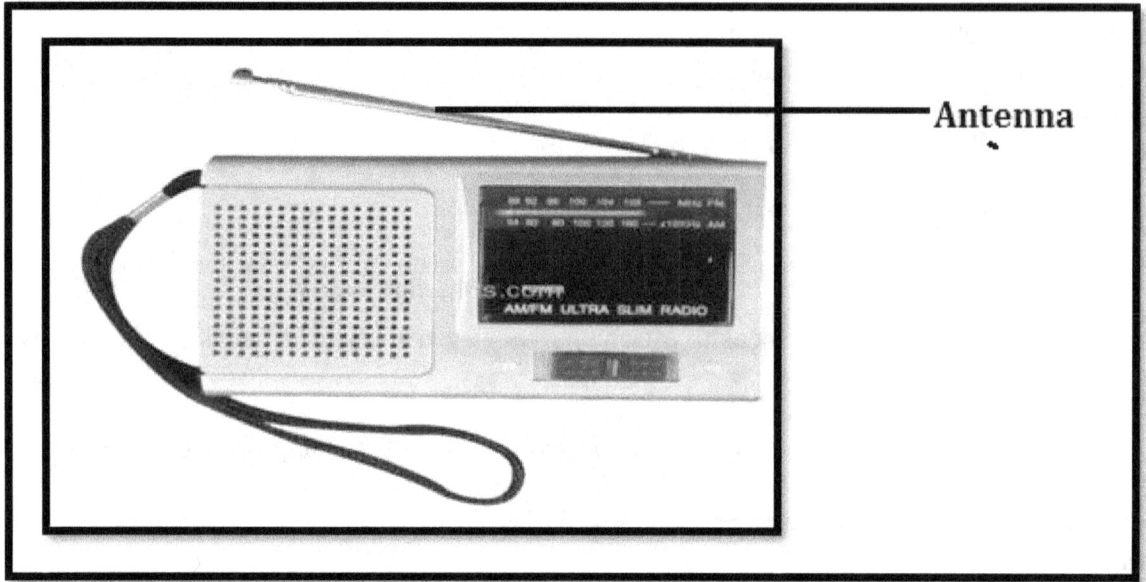

A Simple Radio with AM and FM

Radio receives many different signals. The tuner is adjusted to select the wavelength of the broadcast that is required. The signal is strengthened (amplified) and a loudspeaker turns it into a sound that can be heard.

Broadcasting Radio Waves

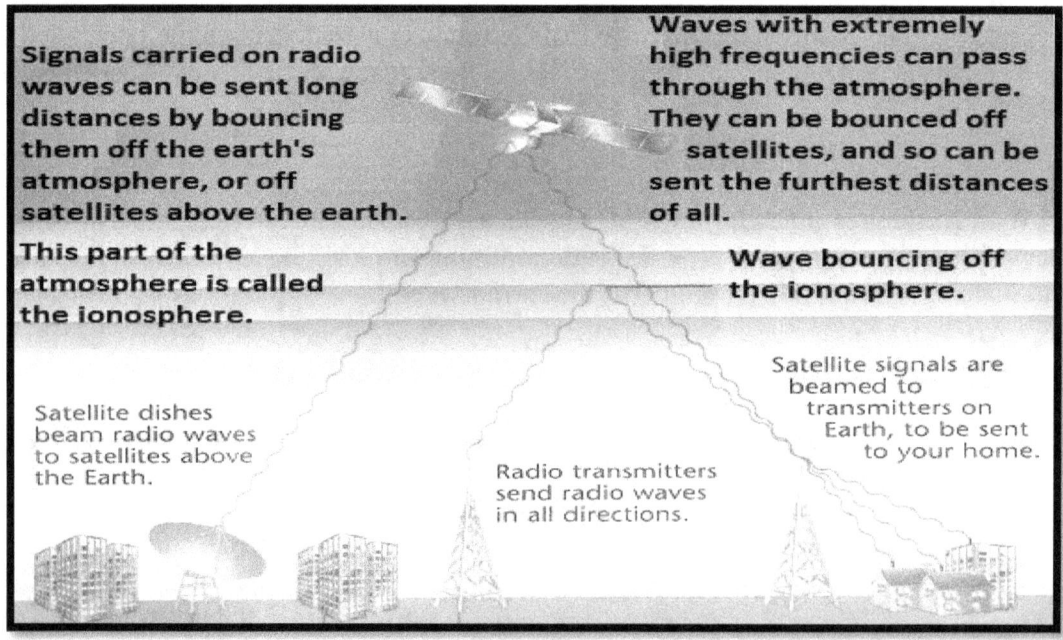

The diagram shows radio waves

Television

Television means to see from afar. Seeing far or nearby requires light. Light forms scenes you see on the television screen. But it is not the light of the original scene, in television, images and sounds travel electronically, that is, by means of electrical energy.

A television camera records the pictures while a microphone records the sounds. The sound and light signals are changed into electrical signals. These signals are sent to the television set (receiver) which converts the electrical signals back to light and sound to give the appropriate picture and sound.

Televisions of the Past

Television of the Present

Cathode Ray Tube (CRT) Television

Flat Screen Television

A television camera changes the light that is reflected from a scene into electronic signals. A device called a transmitter sends out the signals (along with signals for the accompanying sound, which has been picked up by a microphone). Finally, a television receives the signals and changes them back into sound and picture images. Television signals began with a television camera. The television camera has lenses that concentrate light to form images of objects.

The Computer

A computer is a device that performs four functions:

- **Inputs data** -getting information into the machine
- **Stores data**- holding the information before and after processing
- **Processes data**- performing prescribed mathematical and logical operations on the information at high speed
- **O**utputs data- sending the results out to the user via some display method.

  Messages are transformed into signals in one device to be sent, then interpreted by another device.

Parts of A Computer

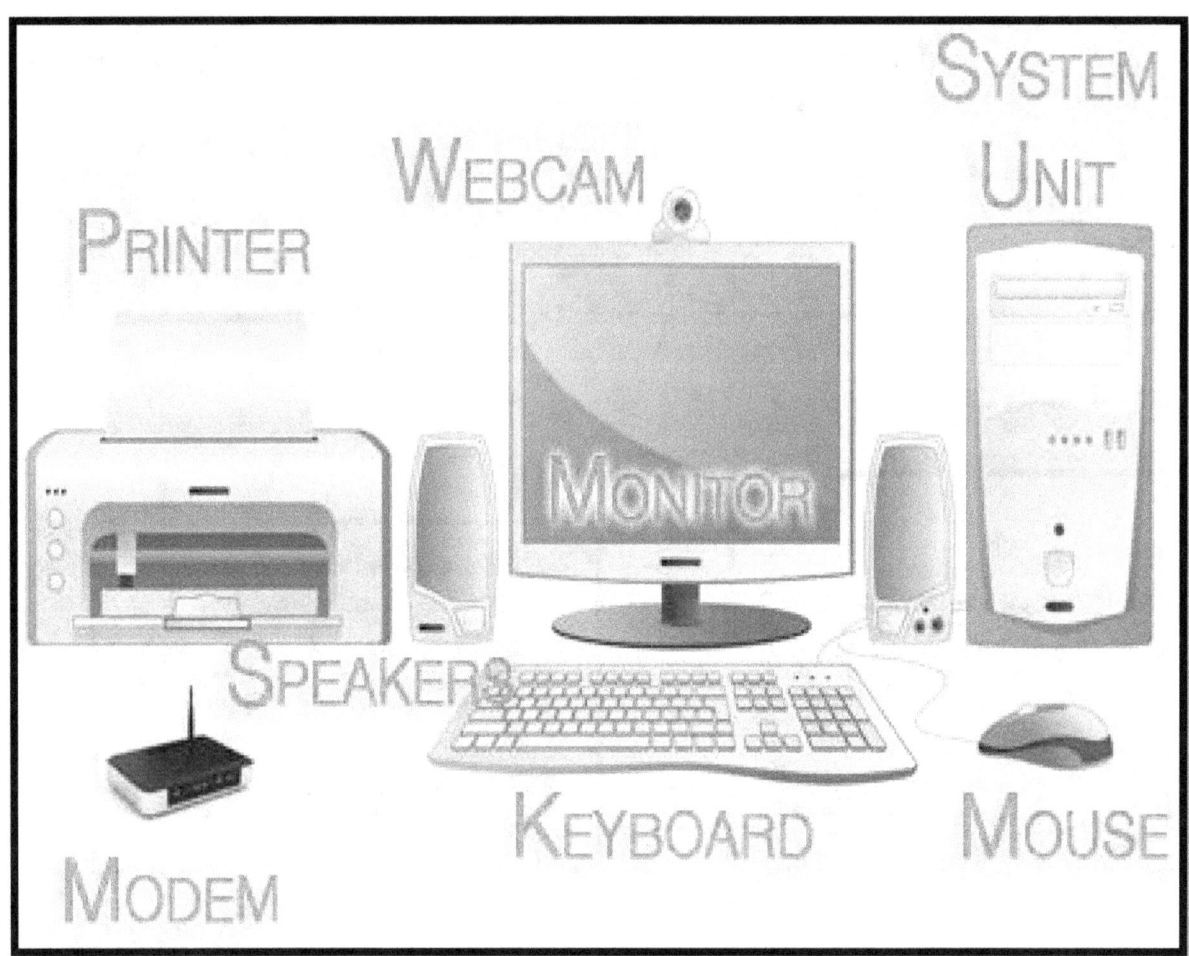

The diagram shows the parts of the computer

Input and Output Devices

The parts of the computer can be summed up as **input** or **output devices.**

As a user, you will interact with the programs running on your computer through the input devices connected to it, such as a mouse and a keyboard. You use these devices to provide input (such as the text of a report you are working on) and also to give commands to the program

The computer program will provide output like the data resulting from the manipulations within the computer via various output devices for presenting the information. Examples of output devices are a monitor, a printer, or a sound output system that beeps if the program needs your attention.

Chart of Input and Output Devices

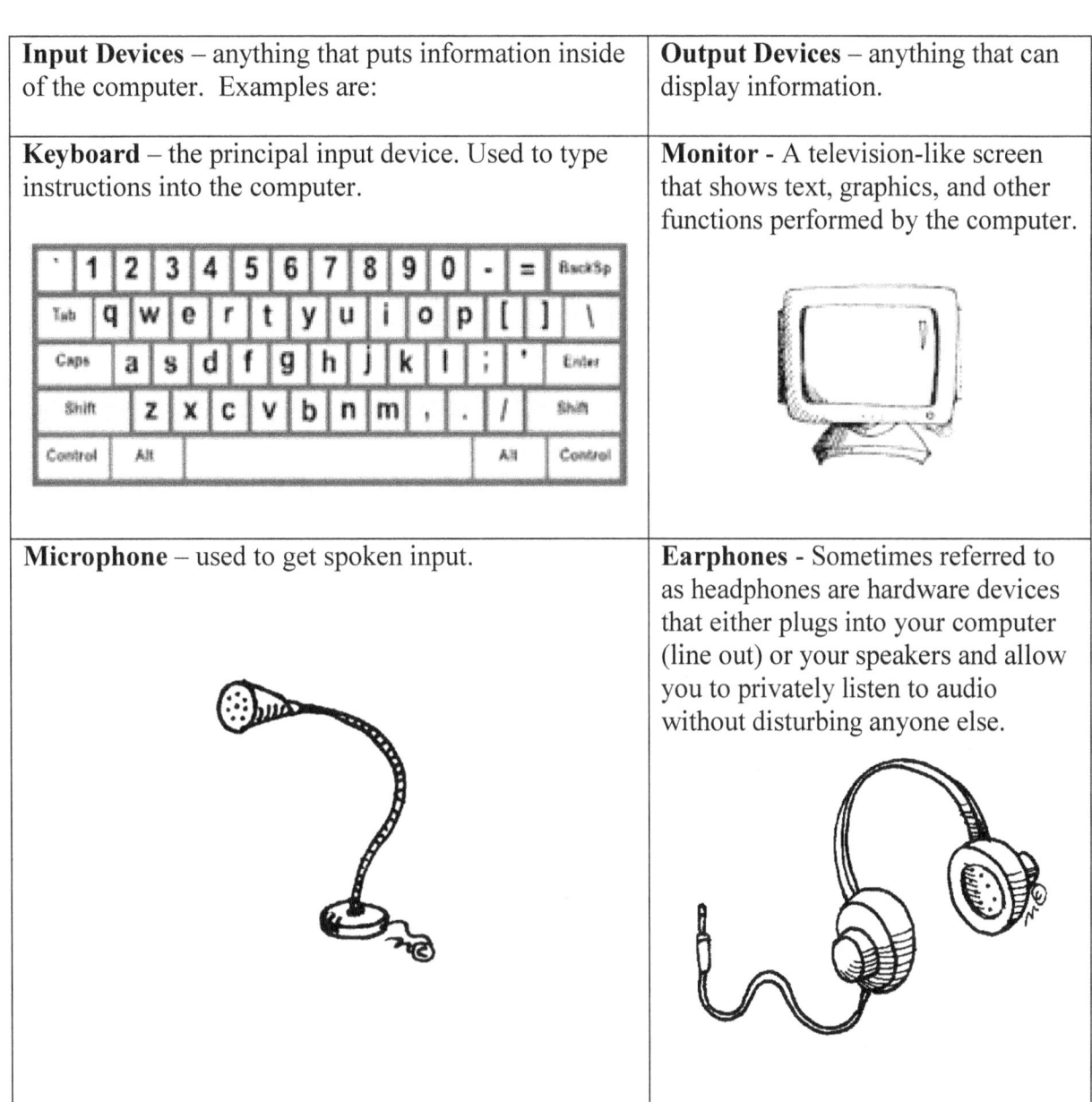

Input Devices – anything that puts information inside of the computer. Examples are:	**Output Devices** – anything that can display information.
Keyboard – the principal input device. Used to type instructions into the computer.	**Monitor** - A television-like screen that shows text, graphics, and other functions performed by the computer.
Microphone – used to get spoken input.	**Earphones** - Sometimes referred to as headphones are hardware devices that either plugs into your computer (line out) or your speakers and allow you to privately listen to audio without disturbing anyone else.

Input Devices (Cont'd)	Output Devices (Cont'd)
Webcam - a video camera that feeds its image in real time to a computer or computer network.	**Printer**- An output device that converts the coded information from the processor into a readable form on paper.
Mouse- A hand-operated electronic device used to move a cursor or pointer on the display screen. Mostly used with microcomputers. When you move the mouse and click, it tells the computer what to do.	**Speakers**- used to produce audio output.
Scanner - a device that optically scans images, printed text, handwriting, or an object, and converts them to a digital image.	**Disk Drives** - A disk drive is used to record information from the computer onto a floppy disk or CD.
Drawing Tablet - similar to a white board, except you use a special pen to write on it and it's connected to the computer. Then the word or image you draw can be saved on the computer.	

The **modem** is the piece of equipment that allows a **computer** or fax to communicate with other **computers** by sending and receiving information over telephone lines. For this reason it is considered both an input and output device.

"Modem" stands for **mo**dulator – **dem**odulator. The modem converts or modulates digital information produced by a computer or fax into analog waves. **Analog** is best explained as the transmission of **signal** such as sound or human speech, over an electrified copper wire. The modem receiving the information demodulates (turns back) the waves into digital code which is understood by another computer or fax.

CPU - Central Processing Unit -The main internal component of a computer where executions of instructions are carried out and calculations are performed

Storage devices - such as disk drives, store your documents and programs when they are not currently in use for processing.
CD-CD-R (Compact Disc - Recordable) and CD-RW (Compact Disk - ReWritable) are CDs that can be written to (if your computer has a CD-RW drive).
DVD- DVD-ROM discs (DVD = Digital Versatile Disc) (ROM) Read-only memory
A USB (Universal Serial Bus) Flash Drive is a portable solid state memory device that plugs into a USB port on your computer.

Hardware and Software

A computer system consists of both **hardware** and **software**. The hardware is the physical equipment: the computer itself and the peripherals connected to it. The peripherals are any devices attached to the computer for purposes of input, output, and storage of data (such as a keyboard, monitor display, or external hard disk).

The software consists of the programs and associated data (information) stored in the computer. A program is a set of instructions that the computer follows to manipulate data. Being able to run different programs is the source of a computer's versatility. Without programs, a computer is just a lot of high-tech hardware that does not do anything. But with the detailed, step-by-step instructions of the program (written by humans) the computer can be used for tasks ranging from word processing a letter to Aunt Mary, to simulating global weather patterns. The computer appears to be so amazing simply because it can execute these sets of instruction *very, very, fast*; but it is just following the program steps one by one in a very simple-minded manner.

How Have Cell Phones Changed Communication?
Mobile Phone

A **mobile phone** (also known as a **cellular phone**, **cell phone**, and a **hand phone**) is a device that can make and receive telephone calls over a radio link while moving around a wide geographic area. It does so by connecting to a cellular network provided by a mobile phone operator, allowing access to the public telephone network.

Cell phones have vastly changed the way we communicate today. A cell phone can be all you need for communicating. From a cell phone you can make calls, text message, BBM message, email, send and receive directions, go on the Internet, buy things, do online banking, listen to music and much more. In that one device, you can do everything. There is no longer a need for multiple devices and you can be on the move (mobile) doing it all.

Significance

The cell phone has become a common product. Almost everyone has a cell phone, and they have only been around for 35 years and popular for about 15. Yet, they have become an indispensable part of daily life. It is hard to imagine what many would do without them.

Function

A cell phone's basic function is to make and receive calls from anywhere you have service. Even though they can do more, calling is their main function. This function enables a person to make or receive a call without having to worry about location. The other basic function is location tracking in case of emergency. By law cell phones are built in with GPS technology that makes them "911 capable." This allows, if the cell phone is powered on, for your location to be tracked.

Types

There are two types of cell phones: regular and Personal Digital Assistants (PDA). Regular cell phones can have similar functions as PDAs, but they are mainly for making and receiving calls. A PDA, personal data assistant, is made more to be a portable office. These devices allow you to have all the functions of your office on the go. With a PDA you can track appointments, go online, email, edit documents, and send business cards.

Benefits

The main benefit of a cell phone is convenience. You have all you need in one device. You no longer need to go out of your way to do something. If you need to make a call on the go, away from home or work, you can. If you need to send an email at lunch you can. If you need directions, you can get them, and if you need to get online, it's there for you. It helps you to get more done.

Warning / Safety

Do not use mobile telephones in open spaces during thunderstorms, while taking baths, at gas stations or while being re-charged. Adhere to safety tips while using a mobile phone and encourage others to do so.

With cell phone popularity rising, there has also been a rise in cell phone related accidents. While being able to use a cell phone in the car is convenient, it has to be safe. Talking and texting while driving is dangerous! If you need to talk while driving there are two options you can choose. You can pull over or buy a hands-free device to use, such as a headset. Additionally, some concern has been raised about cell phone radiation and how it might affect the brain if held against the ear for long periods of time.

Landline vs. Cell Phone

Environmental Concerns
Landlines consume more energy than cell phones, as they remain plugged in at all times. This is true of cordless landline phones as well, because of the charger required. While cell phones generally do not last as long as landlines (as they often become outdated), they are easier to recycle.

Flexibility

An advantage to cell phones is that they can go anywhere. An advantage to land lines is multiple telephone service in different rooms of the house. Land lines with multiple phones can be more cost effective than maintaining multiple cell phones in a household.

Costs

Both landline and cell phone costs depend on a number of factors, including family size, lifestyle and whether long distance calls are made. For large families, it is cheaper to share a single landline than to purchase multiple cell phones, even if they are serviced under a family plan. If an individual's lifestyle involves frequent, long phone conversations, a landline is still the more cost-effective option. This is because landlines have low, single flat rates. When it comes to long distance calls, cell phones may be less expensive (or provide free long distance), but only when used between designated times on designated days.

Contracts

Unlike landlines, most cell phone plans require contracts. There are exceptions, however, as some companies offer pay-as-you-go plans. Generally, if an individual breaks a contract or is late with payments, penalties--in the form of expensive fees--are enforced.

Reliability

While cell phone technology continues to improve, cell phones are still not as reliable as landlines. Especially when it comes to making emergency calls or important career-related calls, landlines are a much more secure option. Cell phones can lose service in some areas and cell phone batteries require recharging and have the potential to die in important situations.

Features

While the landline is more reliable than the cell phone, it lacks the ability to be carried around and utilized in all day-to-day activities. Cell phones also provide callers with the ability to send text messages and, in some cases, take photographs, surf the Internet and play games. With cell phones, wallpapers or backgrounds can be personalized, as well as ring tones.

Reliability
Corded phones operate well in any building except in extreme conditions that result in extended

power outages. Cell phone reliability varies with the service. Most cell phones do not operate well in locations such as tunnels and secured government buildings. However, depending on the reliability of the individual service, cell phones can be used from most locations during an emergency.

Emergency Assistance

Land lines have an advantage over cell phones in terms of 911 emergency service. Emergency dispatchers can determine the originating location of a 911 call placed by a land line telephone. This is not yet the case with cell phones.

Use Patterns

Individual use is a key factor in making an either/or decision. Someone who is rarely at home, such as a frequent traveler, may choose to forgo a home telephone.

Parts of a Mobile Phone

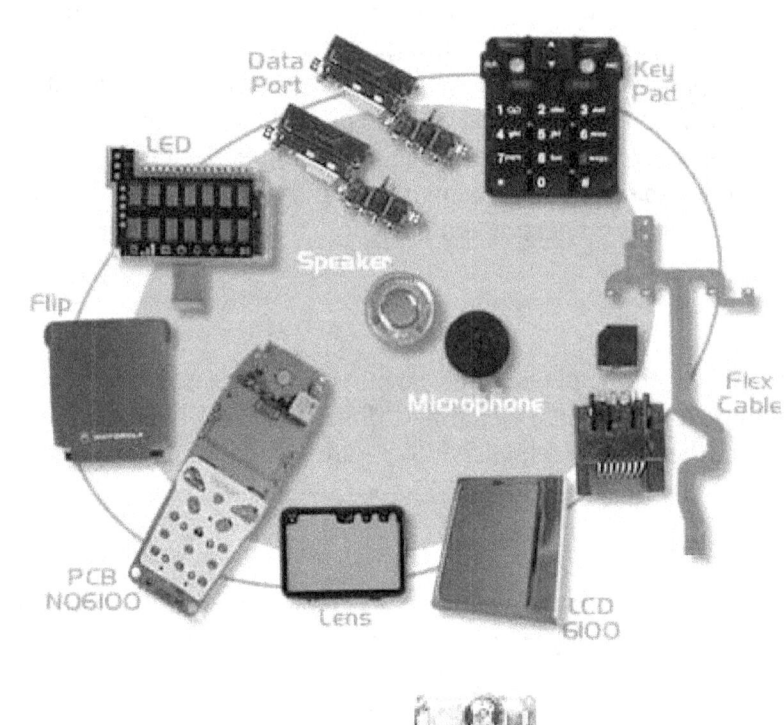

Flex Cable for Motorola   **Speaker**   **LCD Frame**   **LCD Frame Display**

Scientific Principle of the Worldwide Web (WWW)

The Worldwide Web (WWW) and the internet are not one and the same. The internet is a global system of interconnected computer networks. The web is a collection of interconnected documents and other resources linked by hyperlinks and URL's (Universal Resource Locator). The web is an application running on the internet.

Satellite Communications

A satellite is a space vehicle that orbits the Earth which contains one or more radio transponders that receive and retransmit signals to and from the Earth. Satellites allow radio waves that travel in straight lines to be reflected to distant parts of the Earth that would not be accessible by straight lines.

Satellites can be used for communication, research, to forecast weather and for navigation as in the GPS (global positioning system).

The diagram shows the different types of satellite communication systems. The Geostationary orbits (GEO) satellite system is primarily used for television broadcast services, as their satellites appear stationary above the Earth. Medium Earth orbit (MEO) and Low Earth orbiting (LEO) systems are used for mobile communications as they are located much closer to the Earth. However, these satellites continuously move relative to the surface of the Earth.

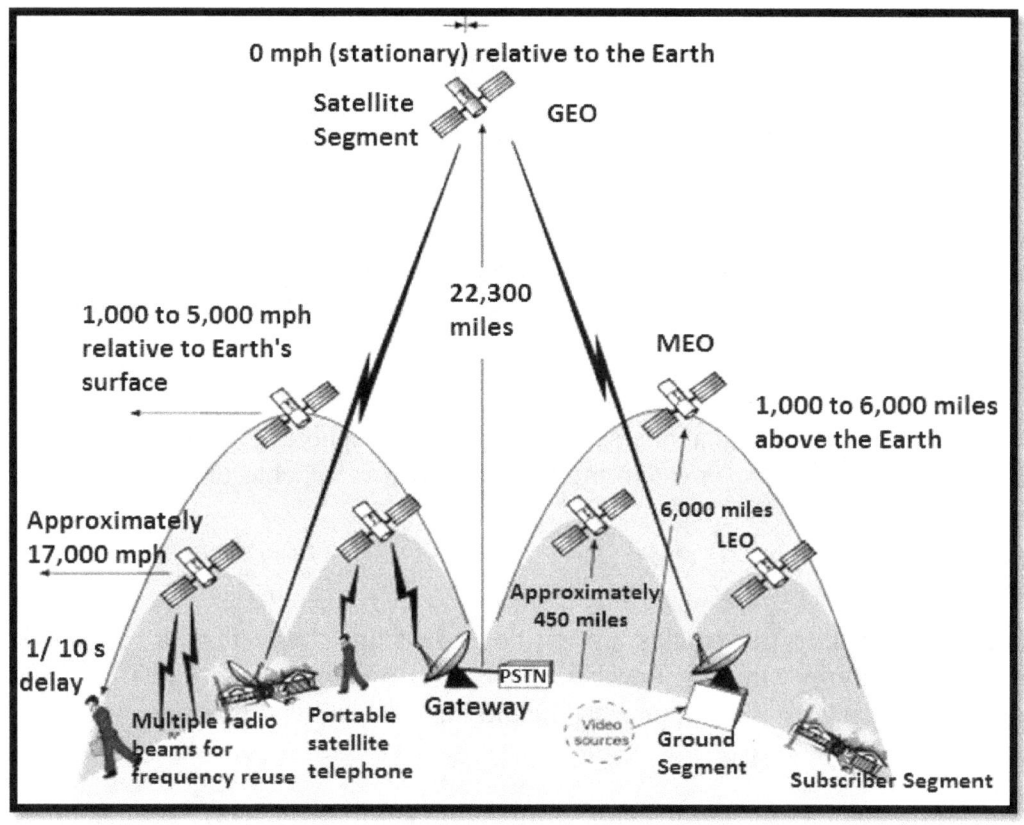

Satellite Communication

- All satellites need to have some means of communication with Earth; the satellite may need to receive instructions and transmit the information it collects, or it may relay information sent to it to another site on Earth
- This is generally done using some type of antenna
- Antennas are defined simply as a piece of equipment that allows transmission and reception of radio signals
- Since the information is transmitted using radio waves, which move at the speed of light, this method allows for very fast communications (only a very small time lag)
- Antennas come in many families: simple, dishes, patch arrays, and inflatable.

The Role of Satellites and Efficiency of Communications

It is difficult to go through a day without using a communications satellite at least once. When you watch T.V., make a long distance phone call, use a cellular phone, a fax machine, a pager, or even listen to the radio, you used a communications satellite, either directly or indirectly.

Communications satellites allow radio, television, and telephone transmissions to be sent live anywhere in the world. Before satellites, transmissions were difficult or impossible at long distances. The signals, which travel in straight lines, could not bend around the round Earth to reach a destination far away. Because satellites are in orbit, the signals can be sent instantaneously into space and then redirected to another satellite or directly to their destination.

Advantages of Satellite Communications

Satellite communications have unique advantages over conventional long distance transmission ns. Satellite links are unaffected by the transmission variations that interfere with hf (**High frequency**) radio. They are also free from the high attenuation of wire or cable facilities and are capable of spanning long distances. The numerous repeater stations required for line-of-sight or tropo-scatter links are no longer needed. They furnish the reliability and flexibility of service that is needed to support a military operation.

Capacity

The present military communications satellite system is capable of communications between backpack, airborne, and shipboard terminals. The system is capable of handling thousands of communications channels.

Reliability

Communications satellite frequencies are not dependent upon reflection or refraction and are affected only slightly by atmospheric phenomena. The reliability of satellite communications systems is limited only by the equipment reliability and the skill of operating and maintenance personnel.

Vulnerability

Destruction of an orbiting vehicle by an enemy is possible. However, destruction of a single communications satellite would be quite difficult and expensive. The cost would be excessive compared to the tactical advantage gained. It would be particularly difficult to destroy an entire multiple-satellite system such as the twenty-six random-orbit satellite system currently in use. The earth terminals offer a more attractive target for physical destruction. These can be protected by the same measures that are taken to protect other vital installations. A high degree of freedom from jamming damage is provided by the highly directional antennas at the earth terminals. The wide bandwidth system that can accommodate sophisticated anti-jam modulation techniques also lessens vulnerability

Activities

1. Describe the technological advances in banking e.g. ATMs, internet / online banking.
2. Observe components of a fax machine: heating elements, scanner, paper entry point, paper exit point, etc.
3. Observe photographs or components of discarded computers. In groups, discuss the methods of communication via computers.
4. Compare the relative efficiency of any of the devices studied: telegraph, radio, telephone, fax machine, mobile phone
5. Some devices convert signals to electricity, light, radio or microwaves for transmission. Classify communication instruments based on their scientific principle.

Index